Critica

General Edit‹

University of L

G000300689

Praise for *The Medium is the Maker*

"Hillis Miller's impact as one of the key theoretical figures of his era is bound to his ability to perpetually re-create his own post-contemporaneity. This is again and quite wonderfully apparent in *The Medium is the Maker* – his cunning engagement with contemporary media, e-culture, and its implications not just for 'literature' but for the epistemo-political orders of life and politics today. At a time when digital-culture critics tend to relapse into precritical technicians of 'new media,' Miller accelerates Derrida's and Freud's errant musings on telepathy to explore today's cognitive centers as traversed by phantom networks (literary and literal), networks out of which worlds are generated, phenomenality programmed, disruption guaranteed." *Tom Cohen, Professor of English at the University at Albany of the State University of New York*

"J. Hillis Miller lulls the reader to a catastrophic awakening to the forms of telepathic transference haunting the forces of legibility. Not so much reading as 'jamming' the voices of Browning, Freud and Derrida, Miller conjures up for us a contemporary big Other whose spectral form is literally pieced together by today's avalanche of bytes and arrays of teletechnic letters. An electrifying performance destined to put the 'flyting' on the psychoanalytic archive." *Sigi Jöttkandt, author of* First Love: A Phenomenology of the One *(2009) and* Acting Beautifully: Henry James and the Ethical Aesthetic *(2005)*

"For decades J. Hillis Miller has been one of the best, most influential, and most prolific literary critics writing in English. Always on the cutting edge of his discipline, Miller has in many of his published works defined the location of that edge for his fellow scholars. In his new book Miller once again takes literary criticism in a fresh and unexpected direction, thinking about the way the most recent communication technologies alter notions of the self, of public and private space, of communication as thought transference, to name only a few of the many issues Miller touches on in this penetrating study. Reading Miller's criticism is always an intellectually rewarding and exhilarating experience, as any reader of *The Medium is the Maker* will attest." *John T. Irwin, Decker Professor in the Humanities, The Johns Hopkins University*

"In *The Medium is the Maker*, Hillis Miller offers an eloquent and entertaining audience with the spirits of three famously telepathic texts. Splendidly didactic, yet as ambitious and creative as it pretends not to be, the book is a fine guide to the turns and transfers of deconstructive reading. There are no jarring notes, but many delicious linguistic surprises in this genial jamming session." *Rachel Bowlby, Northcliffe Professor of English at University College London and author of* Freudian Mythologies: Greek Tragedy and Modern Identities

"This is J. Hillis Miller in sparkling form, showing us why telepathy is a phenomenon we can neither accept nor reject, an affront to our rational selves that reveals something about reason itself. His astute readings of the texts in which three great writers grappled with this irresolvable anomaly lead him to consider the telepathic implications of our new technologies, making of our bodies conduits for messages from around the globe and perhaps beyond it." *Derek Attridge,* University of York, author of *The Singularity of Literature*

"At first glance, J. Hillis Miller's *The Medium is the Maker* succeeds admirably at its self-appointed task: a rich and rigorous updating of Marshall McLuhan's 'The Medium is the Message' in terms of key theaters of contemporary theoretical deliberation, above all with respect to deconstruction, speech-act theory, and rhetorical reading. But the volume is far more than that. Written at the pitch of verve and free-wheeling readerly joy that Miller has attained in other of his tele-technic writings (notably in *Illustration* and *Black Holes*), it is an indispensable meditation on systems of communication in their relations to the interlocking directorates of power. It is also a touchingly frank and open inquiry into informed critique's persistent role as an agent of social awareness and political choice.

The Medium is the Maker takes off from a premise that Miller shares with Jacques Derrida, concerning an irreducible "telepathic" dimensions of novels and other literary genres and speech-act situations. Bringing his vast knowledge of Victorian and twentieth-century letters to bear on the question, Miller deftly pursues a surprising constellation of telepathic phenomena articulated by authors as diverse as Robert Browning, Sigmund Freud, and Derrida himself. He focuses on Browning's exaggerated antagonism toward a real-life medium of his day; Freud's struggle with the inside/outside mind-games that are part and parcel of the transference-relationship, and Derrida's recourse to post-cards and other personal forms of address in process of elaborating the philosophical conditions of language and its artifacts. It emerges in the course of Miller's invariably lively elucidations of these three major innovators' work that the metaphysics of telepathy plays a pivotal ongoing role in our interaction with such contemporary media as the Worldwide Web, Internet, and cell-phones. At the same time that he agonizes over the potential negative impact of such media on the institution of literature and *belles letters* in general, Miller forges bravely into this new world of mediation, affirming that lucid, free-ranging, and inventive critique, which his commentary exemplifies, is as indispensable as ever.

This terse and slender volume will richly reward an audience vastly larger than the usual academic suspects. It deserves a boisterous main-stream audience indeed!" *Henry Sussman, Department of Germanic Languages and Literatures, Yale University*

For Nicholas Royle

The Medium is the Maker

Browning, Freud, Derrida and the
New Telepathic Ecotechnologies

J. Hillis Miller

sussex
ACADEMIC
PRESS

Brighton • Portland

2 4 6 8 10 9 7 5 3 1

First published 2009 in Great Britain by
SUSSEX ACADEMIC PRESS
PO Box 139   Eastbourne BN24 9BP

and in the United States of America by
SUSSEX ACADEMIC PRESS
920 NE 58th Ave   Suite 300
Portland, Oregon 97213–3786

British Library Cataloguing in Publication Data
A CIP catalogue record for this book is available from the British Library.

Library of Congress Cataloging-in-Publication Data
Miller, J. Hillis (Joseph Hillis), 1928–
The medium is the maker : Browning, Freud, Derrida, and the
  new telepathic ecotechnologies / J. Hillis Miller.
p. cm.
Includes bibliographical references and index.
ISBN 978-1-84519-319-5 (p/b : alk. paper)
1. Communication in literature. 2. Telepathy in literature.
3. Mediums in literature. 4. Psychoanalysis and literature.
I. Title.
PN56.C662M43 2009
809'.93356—dc22

2009019450

The paper used in this book is certified by
The Forest Stewardship Council (FSC).

Typeset and designed by SAP, Brighton & Eastbourne.
Printed by TJ International, Padstow, Cornwall.
This book is printed on acid-free paper.

Contents

Series Editor's Preface

When reading this book look out for the author; he's the one who goes by the name 'I,' 'me,' 'myself,' 'yours truly,' but most often by his initials 'JHM.' And I would not trust him if I were you for he will quote a dead man such as Sigmund Freud, Robert Browning or, his great friend, Jacques Derrida, only then to interrupt them to tell us what the dead man really is trying to say. So, he's like a medium; and, as I say, he is a medium who signs himself 'JHM'— again and again he marks his mediations with this monogram. And I believe it is a kind of code, or clue to what is, as you will see, the central question in this book; this question is, to quote JHM (prepare yourself): 'Why all this jam?'

Why indeed? Well, JHM says it is all to do with how Freud makes several passing references to jam in his writing. JHM suggests we think about all this jam in relation to *Through the Looking Glass* where the White Queen declares, 'The rule is, jam to-morrow, and jam yesterday—but never jam today.' JHM says this is the rule of communication, since (he says) any form of communication, even speaking (my voice must travel through air and thus time before you hear me) involves some kind of delay that makes impossible the presence of the present, the now. So, no jam today—never.

All this is very intriguing, but still more intriguing is the coded message that the author so deftly leaves in the form of his monogram, 'JHM,' which is, of course, so close to (yes, you saw it coming) 'JAM.' Now, the coded message (which I have cracked) is this: that JAM is all about 'yours truly,' 'myself,' 'me,' 'I,' the first person singular—for 'Jam' is, you see, a conflated, squeezed and inter-linguistic (JHM is keen on such) version of that most fundamental phrase of all time and all places: namely, 'I am.' You see,

'Jam' is really 'Je am' or 'I am' which, of course, in French, in the language of Jacques Derrida, is 'Je suis' which, as JD himself reminded us, can mean not only 'I am' but also 'I follow' which is particularly uncanny given that 'Je suis' (as JD also reminded us) is mighty close to the name of the one who, some say, laid claim to the divine title, 'I Am'—and he, of course, was that fellow Jesus.

You may shake your head, but this book is all about belief—technically about belief in telepathy ('Do I believe in telepathy?' asks JHM) but the truth is that it is really about belief per se. As JHM reminds us, Derrida says that 'it is impossible not to believe'—again he is, ostensibly, referring to belief in telepathy, but when I read this book it becomes clear to me that what the now-dead man really means to say is that it is impossible not to believe *in believing*.

So: do we believe Jacques Derrida? Would we credit it? Well perhaps we would not credit anything after the recent financial disasters of which JHM is so determined to remind us. As we have learnt, the whole business of credit is an appallingly risky business, but then (listen now): risks *must* be taken. JHM is adamant about this. I know because he here confesses that once he took Derrida sailing on Long Island Sound in a small yacht despite the fact that the '"Small Craft Warnings" were up and we should probably have not been out sailing.' JHM did not tell Derrida about the warnings, they just went to sea, pushed the boat out, and risked death by drowning (an obsession of Derrida's). As JHM admits, he could so easily have been in a jam on dangerous water, all at sea—jam all at sea, *je* am all at sea, *je suis* all at sea . . . Jesus all at sea, and where else? Water was his medium. Don't believe me. Read for yourself.

SJS
Lancaster, February 2009

The Critical Inventions Series

Do I dare / Disturb the universe?
(T. S. Eliot, 'The Love Song of J. Alfred Prufrock', 1917)

In 1961 C. S. Lewis published *An Experiment in Criticism*; over forty years later, at the beginning of a new century, there is pressing need for a renewed sense of experiment, or invention in criticism. The energies unleashed by the theoretical movements of the 1970s and 1980s have been largely exhausted—many now say we are experiencing life after theory; some, indeed, say we are experiencing life after criticism. Criticism, we might say, is in crisis. But that is where it should be; the word 'criticism' comes, as we know, from the word 'crisis'.

Talk of crisis does not, though, fit easily within the well-managed contemporary academy; with its confident talk of 'scholarly excellence', there is a presumption that we all know, and are agreed upon, what scholarship and criticism is. However, to echo Paul de Man, 'we don't even know what reading is'; and what is, potentially, exciting about our present crisis is that now we really know that we don't know what reading is. It is, then, in a spirit of learned ignorance that we propose 'critical inventions', a series which will feature books that, in one way or another, push the generic conventions of literary criticism to breaking point. In so doing the very figure of the critic will shift and change. We shall, no doubt, glimpse something of what Oscar Wilde famously called 'the critic as artist', or what Terry Eagleton called 'the critic as clown'; we may even glimpse still more unfamiliar figures—the critic as, for example, autobiographer, novelist, mourner, poet, parodist, detective, dreamer, diarist, flaneûr, surrealist, priest, montagist, gambler, traveller, beggar, anarchist . . . or even amateur. In short, this series seeks the truly critical critic—or, to be paradoxical, the critic as critic; the critic who is a critic of criticism as conventionally understood, or misunderstood. He or she is the critic who will dare to disturb the universe, or at least the university—in particular, the institutionalisation of criticism that is professional, university English.

Establishment English is, though, a strange institution that is capable of disestablishing itself, if only because it houses the still stranger institution of literature—which, as Jacques Derrida once wrote, 'in principle allows us to say everything/ anything [tout dire]'. We, therefore, do not or cannot yet know of what criticism may yet be capable—capable of being, capable of doing. 'Critical inventions' will be a series that seeks to find out.

> Read the text right and emancipate the world.
> (Robert Browning, 'Bishop Bloughram's Apology', 1855)

JOHN SCHAD, Series Editor

JOHN SCHAD is Professor of Modern Literature at Lancaster University. He is the author of *The Reader in Dickensian Mirrors*, *Victorians in Theory*, *Arthur Hugh Clough*, and *Queer Fish: Christian Unreason from Darwin to Joyce* —this last published by Sussex Academic Press. He is also the editor of *Dickens Refigured*, *Thomas Hardy's A Laodicean*, *Writing the Bodies of Christ* and co-editor of *life.after.theory*.

Our world is the world of the "technical," a world whose cosmos, nature, gods, entire system, is, in its inner joints, exposed as "technical": the world of an *ecotechnical*. The ecotechnical functions with technical apparatuses, to which our every part is connected. But what it *makes* are our bodies, which it brings into the world and links to the system, thereby creating our bodies as more visible, more proliferating, more polymorphic, more compressed, more "amassed" and "zoned" than ever before. Through the creation of bodies the ecotechnical has the *sense* that we vainly seek in the remains of the sky or the spirit. (Jean-Luc Nancy, *Corpus*)

O FTEN I ask myself: how are fortune-telling books [in English in the original], for example, the Oxford one, just like fortune-tellings, clairvoyants, mediums [*tels les dits de bonne aventure, les voyants ou les médiums*], able to form part of what they declare, predict, or say they foresee even though, participating in the thing, they also provoke it, let themselves at least be provoked to the provocation of it? (Jacques Derrida, "Telepathy")

The Boomerang Effect

N OT Marshall McLuhan's "The medium is the message," but a new formula: "The medium is the maker." *Poeisis* means "making." The poet is a maker. Usually, people once thought, the poet is a maker of beautiful lies. But "maker" also implies a performative force. The medium itself makes something happen. Lies, if they are believed, bring something about, such as the invasion and occupation of Iraq. This happening, however, occurs, according to Derrida, without reference to the message the medium carries. He says this, for example, about a postcard you happen to intercept: "So then, you commit yourself and you commit your life to the program of the letter, or rather of a postcard, of a letter that is open, divisible, at once transparent and encrypted. The program says nothing, it neither announces nor states anything, not the slightest content [*pas le moindre contenu*], it doesn't even present itself as a program. One cannot even say that it 'makes like' a program [*qu'il 'fait' programme*], in the sense of appearance, but, without seeming to, *it makes* [il fait], it programs."[1] The force of an annunciation that announces nothing acts whether the medium is the speaking voice, often a voice speaking in ventriloquism for some other or other, or coded table-rapping, or the printed word, or hand-written postcards and letters, or telephone interchanges, or cinema, or television, or computer files, faxes, emails, instant messaging, texting (txtng)

I

with iPhones or Blackberries,[2] or blogs on the Internet. It does not matter what message you try to convey through one or another of these media. The medium works on its own, performatively, in a unique way in each case, to bring something about, to make something happen. It makes.

This essay is no more than an extended footnote to Nicholas Royle's brilliant *Telepathy and Literature: Essays on the Reading Mind* and to his other writings on telepathy.[3] I am especially indebted to Royle's claim that the "omniscient narrator" in realist fiction would be better defined as a telepathic medium who (or which) has terrifying clairvoyance about what the characters are thinking and feeling. "Telepathy," he notes, entered the English language only in 1882. Earlier it was called "clairvoyance" or "sympathetic clairvoyance."[4] This book is also written in homage to Royle's admirable translation of Derrida's "Telepathy."

After Royle's work on telepathy, Pamela Thurschwell's,[5] Martin McQuillan's,[6] Marc Redfield's,[7] Julian Wolfreys',[8] and others, all in the wake of Freud and Derrida's telepathy essays,[9] what more is there to say? I mean, what is there left to say, these days, about telepathy, especially now that the new telecommunications media—telephone, radio, tape recorders, cellphones, TV, the Internet, email, instant messaging, blogs, and so on—have made more or less instantaneous touching-, feeling-, knowing-, seeing-, hearing-at-a-distance the most everyday experience imaginable? "Reach out and touch someone" was the old slogan of AT&T. All these new telecoms may explain why experiments to prove or disprove telepathy are now, to a considerable degree, out of fashion. Derrida, in "Telepathy," mentions in 1979 "successful experiments the Russians and Americans are doing with their astronauts" (Te, 236; Tf, 247). These experiments, implies the one or another of Derrida's personae who is speaking, scientifically prove that telepathy works. I do not hear much, these days, about such experiments. We have telepathy as an ordinary part of our lives, so spiritualism proper does not concern us all that much.

Our everyday telepathy, however, is significantly different from the one Freud had in mind, that is, one person's direct clairvoyant knowledge of what someone else is thinking or feeling. Our forms of telecom telepathy give us hearing and seeing at a distance, as in television news, but not access at a distance to the mind of another, such as telepathic narrators in realist fiction grant in imagination to their readers. George Eliot dramatized such access, in *The Lifted Veil* (written 1859; published 1878), as a disaster if it were to occur in real life. This, she thought, is because we would then be able to have insight into the shallow, egotistical, selfish, or even criminal thoughts of our neighbors and loved ones. Modernist and, especially, postmodernist fiction make less and less use of the telepathic narrator, though it is of course still often employed. Post-modern novels often imprison the reader in ambiguous appearances, just as does television. We can see and hear, but not penetrate within. We have no way of knowing just what was going on in General Petraeus's mind when he testified in spring 2008, before Congress and the media, that things are going pretty well in Iraq, thanks to the "surge," and that we need only keep sufficient troops there indefinitely to achieve "victory." Did he know he was lying, or, to put it more delicately, stretching things quite a bit? No way to know.

Enough interest remained in spiritualism not long ago, however, for Gian Carlo Menottti to have written an opera called *The Medium* (1946), as well as one called *The Telephone, or L'Amour à trois* (1947). The premier of *The Telephone* was presented in a double bill with *The Medium* on February 18–20, 1947. *The Medium* is about a fraudulent medium who is driven into a murderous frenzy by what she fears is a genuine supernatural manifestation. *The Telephone* is about a young man who cannot propose to his beloved in person because she is always so busy talking on the telephone. The telephone is the third in this techno-amorous triangle. The young man finally calls his beloved from a telephone booth and successfully proposes. Even marriage proposals must be mediated these days by some technological medium.[10] Think of all those telephone booths in Derrida's "Envois," in *The Post Card*! Today, telephone booths have more or less vanished. They have been rendered obsolete by the cellphone. Superman, these days, as a recent *New Yorker* cartoon observed, would have difficulty finding a telephone booth in which to change costumes and selves. The telephone booth used to be the place of exchange and transference, but no more.

What is the connection between Menotti's two operas? They are both, in different ways, about telepathy. Freud, in the amazing last paragraph of "Dreams and Occultism," from the *New Introductory Lectures on Psycho-Analysis*, defines telepathy as probably open to a scientific explanation, just as is talking on the telephone. The telephone is a telepathic device. "The telepathic process," says Freud, "is supposed to consist in a mental act in one person instigating the same mental act in another person. What lies between these two mental acts may easily be a physical process into which the mental process is transformed at one end and which is transformed back once more into the same mental one at the other end. The analogy with other transformations, such as occur in speaking and hearing by telephone, would then be unmistakable. And only think if one could get hold of this physical equivalent of the psychical act!" (SE 22: 55).

Only think! The telephone, in those days, worked through wires, though Freud once, in a passage cited by Derrida, defines telepathy as "a psychical counterpart to wireless telegraphy" (*gewissermassen ein psychisches Gegenstück zur drahtlosen Telegraphie*) (Te, 259; Tf, 269). Telegraphy *sans fil* appears much more occult than the wired kind. Nevertheless, we have rapidly become accustomed to cellphones. How does my voice travel invisibly and inaudibly through the air? It must be by some kind of magic, some form of ecotechnical telepathy, mingling the technological and the natural, the ecological environment. We might also think of instant messaging, in which thoughts flow through the fingers into letters on the computer keyboard and are then are transmitted by a whole series of digital relays, including wireless ones, satellites, and so on, to their preordained destination. They reach their goals, that is, if surveillance techniques or "spam filters" do

4

not intercept them on the way, in our present-day form of what Derrida called *destinerrance*, that is, the fateful deviation of a message from its intended recipient. A given message forms a tiny part of the immense Internet system. Ultimately, but nevertheless almost instantaneously, after many rapid relays and transferences, the letters I type on my keyboard appear again on the addressee's screen as words he or she then translates spontaneously back into thoughts.

Derrida's exuberantly hyperbolic response, in "Telepathy," to Freud's "only think," is to imagine our enforced participation in a gigantic frightening system of irresistible telepathic transfers. Each of us is perpetually "plugged in" to this system, willy nilly. No more privacy:

> But once again, a terrifying telephone (and he, the old man [Freud], is frightened, me too); with the telepathic transfer, one could not be sure of being able to cut (no need now to say *hold on* [in English in the original], *don't cut*, it is connected [*branché*] day and night, can't you just picture us?) or to isolate the lines. All love would be capitalized and dispatched by a central computer like the Plato terminal produced by Control Data: one day I spoke to you about the Honeywell-Bull software called Socrates, well, I've just discovered Plato. (I'm not making anything up, it's in America, Plato.) So he is frightened, and rightly so [I don't see that Freud is frightened. Rather, he seems exhilarated by the power that control over telepathy might give. "Only think," he says, if we could harness and control that power. It is Derrida (and I) who are frightened. (JHM)], of what would happen if one could make oneself master and possessor (*habhaft*) of this physical equivalent of the psychic act, in other words (but this is what is happening, and psychoanalysis is not simply out of the loop [*hors du coup*], especially not in its indestructible

[*increvable*] hypnotic tradition), if one had at one's
disposal a *teknē telepathikē*. (Te, 242; Tf, 252–3)

Derrida rightly observes, on the same page of "Telepathy," that
the analogy Freud draws between telepathy and the telephone is an
example of the ubiquitous dependence of his discourse on various
forms of transference, in many senses, including the "Freudian"
one: "the telepathic process would be physical in itself, except at
its two extremes; one extreme is reconverted (*sich wieder umsetz*) into
the psychical same at the other extreme. Therefore, the 'analogy'
with other 'transpositions,' other 'conversions' (*Umsetzungen*),
would be indisputable, for example, the analogy with 'speaking
and listening on the telephone.' Between the rhetoric and the
psycho-physical relation, within each one and from one to the
other, there is only translation (*Übersetzung*), metaphor (*Übertragung*), 'transfers,' 'transpositions,' analogical conversions, and
above all transfers of transfers: *über, meta, tele, . . . trans*" (Te, 242;
Tf, 252).

This recognition of the ubiquity of transfers when we use any
telecom medium is important for my strategy in this book, as it
moves, by a series of back and forth transfers, from Derrida and
Freud, to Robert Browning, to Daniel Dunglass Home, to
Browning's poem, "Mr. Sludge, 'The Medium,'" back again to
Freud and Derrida, to email on the Internet, to Matthew Freud and
Elisabeth Murdoch, to yours truly as mediator, there from the
beginning and throughout, as the switching point, medium, or
what is still called the telephone "operator," the support, *khora*, or
subjectile, of all these transfers and transpositions.

One perhaps unintentionally comic analogy or transfer Freud
makes is a reference to the way insects seem to be able to communicate with one another by way of an occult telegraphy: "It is a
familiar fact that we do not know how the common purpose comes
about in the great insect communities: possibly it is done by means
of a direct psychical transference of this kind" (SE 22: 55).[11]
Possibly, though scientists now claim to have figured out that ants

6

and bees communicate by sign language. Bees speak by wingbeats and "dances." Ants speak by chemicals called pheromones that are excreted by a given ant and then smelled by the "antennae" of other ants. "Antennae" is a significant word in our days of wireless communication! Ants' antennae are their noses. Ants smell signs, whereas we hear or see them, for the most part. Ant pheromone language has at least twenty words, twenty different kinds of chemical signs. One example is a "follow me" sign that is laid down as a trail on the ground by an ant that has found food.[12]

The indubitable (according to Freud) existence of the unconscious in human beings is a crucial part of his theory of dreams and of telepathy. Do ants and bees have an unconscious? If not, the analogy of putative human telepathy with what he imagined ant telepathy to be like hardly holds. If ants do have an unconscious, this leads to the ludicrous image of Freud psychoanalyzing an ant stretched out on his famous couch: "Tell me your dreams, especially the telepathic ones."

The computer devices or programs called Plato and Socrates have long since become obsolete. The names fascinated Derrida because "Envois," of which "Telepathy" was a misplaced, "transferred," segment, was instigated by a postcard from the Bodleian Library reproducing an illustration from a thirteenth-century fortune-telling book by Matthew Paris. The illustration, absurdly, shows Plato dictating as Socrates writes, or, perhaps, erases. "Envois," inexhaustibly, presents different incompatible and often comic, or even obscene, readings of this postcard. Plato and Socrates hardware and software are long gone, though Honeywell still exists. It is a maker of thermostats, electric heaters, and other such feedback devices. Control Data survives in the ghostly form of the companies into which it was dissolved by mergers. These include the huge financial services company Citigroup, one of the culprits in the financial meltdown that has brought about a global recession.

Sic transit, though much could be said about the dependence of present-day financial "institutions" and transactions on computer transfers. No computers, no Citigroup. And no subprime mortgage catastrophe and the consequent meltdown of the global financial system, requiring what some analysts say amounts to a seven trillion dollar bailout by the American taxpayer, if risky loans the government is taking over are counted in. The foolish greed of the investment bankers has greatly contributed to this economic "state of exception." Our gift of billions of dollars has so far (February 2, 2009) had little effect. The banks have just squirreled it away, or used it to pay themselves big bonuses, or to buy other banks. Derrida would have called their behavior throughout a confirmation of his claim that all social entities are undermined by an irresistible penchant toward autoimmune self-destruction. As Nicholas Royle observes in an email to me, "Ah, credit and credibility: one has only to believe, as 'Telepathy' tells us."

The subprime débâcle occurred by way of complex bundled unsafe mortgages called "derivatives," and derivatives of derivatives called "credit default swaps." This was a gigantic Ponzi scheme of transfers and transfers of transfers that magically created apparent wealth out of nothing, or almost nothing: the mortgagees' signatures on the original deceptive mortgages and the putative value of their houses. Once the inflated market value of those houses plummeted, the bubble burst, as was bound to happen sooner or later. Only computers, with their gifts for spectrality, could so rapidly create and maintain such pyramidal airy nothings.[13] The media, however, especially financial news programs on CNBC, Fox Business, Bloomberg TV, and PBS's *Nightly Business Report*, helped greatly by giving no advance warning of the looming financial tsunami. Deceptive television advertisements for credit cards and mortgage loans have also helped.[14]

Central computers that control everything and know everything, even emailed love letters, are much more with us today than they were when Derrida in 1979 prophetically foresaw what would happen. Illegal and unconstitutional electronic surveillance of any United States citizen is not only possible. It was practiced big time under George W. Bush's presidency, with the connivance of the telecom industry. Even what I write inside my computer, by way of my word processing program, such as this book that I am writing at this moment, is easily accessible to spying government hackers, assuming my computer is connected to the Internet, which it is. I must assume that whatever I write may appear some day on the front page of the *New York Times*, if anyone cares enough about it to purloin my letters. That is unlikely, but not technically impossible.

We know, if we think about it, that our bodies, our nervous systems, and our brains are penetrated, saturated, inundated, soaked, pierced through and through, at every moment, by an enormous cacophony of invisible electro-magnetic waves resonating at many frequencies and coming from an unimaginable number of different broadcasting sources. It is only a matter of getting the right tuning apparatus to filter all these signals into a single chosen one, leaving out the parasites. You only need the prosthesis of a satellite dish, or a TV set with antenna, or an iPhone, or an FM or AM radio, or a cable connection, or a tape-recorder and playback machine, or a computer that can pick up wireless signals from the router in your home, office, or hotel room, and, *voilà*, the old dream of telepathy, of getting in touch at a distance, is fulfilled in the most hyperbolic way. We can hear today much more precisely what Browning's Mr. Sludge claimed to be able to transmit, that is, the voices of the dead—on an old Beatles record, or on a Luciano Pavarotti or Glenn Gould CD, or on a tape of Jacques Derrida reading "Circonfession," or being interviewed at Loughborough University. I have both of the latter, but cannot bring myself to listen to them. They would prove to me that Derrida is really dead. A CD exists that includes a brief early recording of Robert Browning reading "How They Brought the Good News from Ghent to Aix."[15] You can also hear the bugle call used at the charge of the Light Brigade, recorded on a gramophone in 1890, for free online.[16] Tennyson reading from his poems is available on an audiocassette.[17] Spooky! I found the website reference to the Browning CD in two minutes by way of Google, just as I found the passage about the gramophone in Joyce's *Ulysses*, cited in footnote 14, in the same way and at the same speed. No need for a printed copy of *Ulysses*. The Internet swarms with recorded ghosts.

The deep connection of the telephone with death messages and voices from the dead is strikingly exemplified not only in the episode of the grandmother's death in Proust's *À la recherche du temps perdu* (when Marcel talks to her on the telephone, it seems to him she is already virtually dead), but also in the report of the way Mary Baker Eddy, the founder of Christian Science, had a telephone with an outside line placed in her casket in case she was not really dead and needed to call for help, as well as in a news story I heard and saw just today, as I am doing the final revisions of this book (February 5, 2009). This was a segment of the six o'clock television local news from the NBC station in Bangor, Maine, WLBZ2. It seems that many people, in Maine at least, are burying their loved ones with fully charged mobile phones placed in the casket with the corpse. One woman had her father's mobile phone number carved on his tombstone, presumably so casual passersby could call him up. That is really weird! Nevertheless, it obeys the profound logic that connects the telephone with death, as Freud's telepathy essays also demonstrate.

This present book is possible only because it is being written on a computer connected to the Internet. This is so because I am exploiting all the prestidigitizing powers of the word processor—ease of virtually limitless revision and interpolation, spell and grammar checks, possibility of multiple backups, ability to send the file anywhere in the world as an attachment to an email, ability to "google" and "wiki" almost anything, and so on. I know Browning is really dead, but the CD that includes his reading serves as a spectral medium to bring his ghost back, dead or alive, or dead/alive, as a revenant. We no longer need mediums in the occultist sense. The new mechanical recording techniques have put mediums like Sludge, and perhaps also psychoanalytic mediums like Freud, out of business. It is not an accident that both a spiritualist gathering and a psychoanalytic session are called *séances* in French.

The word "medium" names, among other things, a man or woman who is the apparatus for downloading voices, or writing, or table-tapping from the spirit world. The emphasis on physical supports, for communications from the dead, is important. To get through from "the other side," the spirits need tables to turn, or slates to write on, or writing paper, or the medium's body, or his or her hands or voice-box, or ectoplasm recorded on a photographic plate, or a ouija board. With all forms of telepathy, traditional, modern, or postmodern, it is always a question of transferring spirit to some form of matter that can then be read as comprehensible signs and turned back into spirit, that is, "meaning." In short, some "medium" is necessary. Robert Browning, in "Mr. Sludge, 'The Medium,'" has Sludge define a medium as a window (another physical image) onto the spirit world and as the indispensable means, though a cloudy one, through which spirits can speak. He is imagining what he might say in explanation to those at a *séance* when he gets something garbled, spells "Bacon," that is, the seventeenth-century Francis Bacon, "Bakyn," for example. What Sludge reports himself as saying is a succinct summary of what nineteenth- and twentieth-century believers in spiritualism hypothesized:

"You see, their world's [that of the spirits] much like a jail broke loose,
While this of ours remains shut, bolted, barred,
With a single window to it. Sludge, our friend,
Serves as this window, whether thin or thick,
Or stained or stainless; he's the medium-pane
Through which, to see us and be seen, they peep:
They crowd each other, hustle for a chance,
Tread on their neighbor's kibes, play tricks enough!
Does Bacon, tired of waiting, swerve aside?
Up in his place jumps Barnum—'I'm your man,
I'll answer you for Bacon!' Try once more!"

Or else it's—"What's a 'medium'? He's a means,

Good, bad, indifferent, still the only means
Spirits can speak by; he may misconceive,
Stutter and stammer,—he's their Sludge and drudge,
Take him or leave him; they must hold their peace,
Or else, put up with having knowledge strained
To half-expression through his ignorance."[18]

The reader will note that Sludge easily quotes Shakespeare. He allows Shakespeare to speak through him in that phrase about "their neighbor's kibes."[19] I shall return to the all-important role of literature in Sludge's discourse. Sludge imagines that the spirits on the other side of the window are eager to communicate with this side. We on this side of the single window are defined as in jail through our exclusion from the omnipresent spirit world, which is "a jail broke loose." The spirits crowd at the window to get in touch with us. Sludge the medium has to have his own built-in tuning apparatus. He needs this to filter out interference, what in French are called *parasites*, and to get on the right frequency, in order to allow one spirit to speak at a time.

Now we no longer need mediums like Sludge. "Medium" now names the whole bio-techno-prosthetic apparatus of one or another of the new means of telecommunications: radio, TV, the Internet, or whatever. We call these "the media," often in suspicious disparagement, on the assumption that the media are purveyors of lies or of ideological mystifications. That is not a bad assumption, but we need to understand that lies and mystifications are built into the media as media, that is, as technical means of reproduction and transmission of what is at a distance.

Some essential distinctions must be made, however. I must, as Wikipedia says, "disambiguate." Browning's Mr. Sludge primarily serves as a medium to establish communication with the ghosts of the already dead, for example Hiram H. Horsefall's dead mother, or Francis Bacon, or Aristotle. P. T. Barnum, however, was still very much alive in 1861, when Browning's poem was probably written. If I understand Browning's lines correctly, Barnum is one of those specters who crowd at Sludge as window, so matters are, as is always the case in this area, not entirely straightforward. Sludge sometimes, for example, uses his "second sight" to prophesy the future.

Freud and Derrida, however, are more interested than Sludge in prophetic telepathic messages or "premonitory dreams" that foretell the future, though some of Freud's examples are of events that are telepathically transmitted at the moment they occur. Our present-day media-mediums claim to put us in touch with people or events that, it is claimed, are taking place right at this moment, as in "instant messaging," or "Live From Lincoln Center" (which may actually describe a pre-recorded performance), or as in a network news broadcast that purports to be "live" (though also often pre-recorded, as in the shots of US soldiers armed to the teeth and breaking down doors in Iraqi homes that are shown, in identical repetition, over and over for months or even years on network TV to accompany the news of the day).

The here and now of television, as Derrida has more than once observed, is a sham, a simulacrum.[20] What *was* there, at some distant here and now, reaches us, in another here and now on the screen, through elaborate delays, relays, and message-shaping filters. The medium is the maker. As the White Queen in *Through the Looking-Glass* says, "The rule is, jam to-morrow and jam yesterday—but never jam *to-day*."[21] We never, with any medium, have an instantaneous presence of the distant present. The medium comes between, and delays, even if only for the fraction of a second, not to speak of the way any medium always reworks that absent present, always turns "facts" into an "artifactualities."

Jam, by the way, figures saliently in Freud's essays on telepathy, whether in the story about how he, Freud, cut his face at the age of two trying to reach something, perhaps jam, on the top of a cupboard (he still has the scar under his beard—see?), or in Freud's fantasy of an improbable quasi-scientific hypothesis: that the center of the earth is made of jam, or in one of his correspondent's dreams of his second wife's newborn twins licking a washbasin of jam clean. The wife is a dream displacement of the correspondent's daughter, just as the whole story may be a displacement of Freud's grief over his daughter Sophie's recent death. Why all this jam? No doubt as a sign of some repressed infantile sexual desire. We adults want jam today, now, this minute, and we can never have it. We can have only perpetually delayed gratification. It's a good thing too, since, as Yeats says, echoing Freud on the deathwish, "what disturbs our blood/Is but its longing for the tomb."[22]

The different orientations among various media toward past, present, or future are not insignificant, though, as I have shown, that orientation is rarely clear-cut in any telepathic record. The change in dominant temporal orientation from past to future to present, however, probably indicates cultural changes over time in the wishes we would like to have fulfilled. The heyday of spiritualism, I might hazard to guess, was a response to the pain of fading religious belief, just as psychoanalysis challenged spiritualism as a way of fulfilling that compensatory function. I thought I had figured that out for myself. I now find, however, that Freud says just that, followed by Nicholas Royle in *After Derrida*. Perhaps I received a telepathic message from the specters of one or the other. "And we must then reflect," writes Freud in "Dreams and Occultism," "that the tradition and sacred books of all peoples are brimful of similar marvelous tales and that the religions base their claim to credibility on precisely such miraculous events and find proof in them of the operation of superhuman powers. That being so, it will be hard for us to avoid a suspicion that the interest in occultism is in fact a religious one and that one of the secret motives of the occultist movement is to come to the help of religion, threatened as it is by the advance of scientific thought" (SE 22: 34).[23]

Nowadays, perhaps, just perhaps, all our new telecommunication devices are prosthetic replacements for religion, for spiritualism, and for psychoanalysis, all three, in a four-stage historical process that of course does not sublate the earlier stages in a definitive *Aufhebung*, but retains each stage throughout, as when Freud compares telepathy to talking on the telephone. People in the nineteenth century could not bear not believing that the dead are not really dead. Analogously, Derrida is right to assert, in "Telepathy," that no theory of the unconscious is easy to conceive without an accompanying belief in telepathy: "Difficult to imagine a theory of what they still call the unconscious without a theory of telepathy. They can be neither confused nor dissociated" (Te, 237; Tf, 247-8). Derrida, in spite of his resolute assertions, in late inter-views and seminars, that each of us is a windowless monad in a world without God,[24] nevertheless, at the same time, in blank contradiction, believed that *le tout autre*, the wholly other, speaks within each of us to make implacable demands. That happens by a species of telepathic instant messaging.

Derrida asserts, in "Telepathy," that it is impossible not to believe in telepathy. It is impossible not to believe that each of us has an internal television screen by means of which we have visions of what distant friends and relations are thinking, or not to believe that whatever we think is broadcast to the internal television screens of others. The figure of contemporary prosthetic telecom-munications devices is crucial to Derrida's formulation:

> The truth, what I always have difficulty getting used to:
> that nontelepathy is possible. Always difficult to imagine
> that one can think something to oneself [*à part soi*], deep
> down inside [*dans son for interieur*], without being
> surprised by the other, without the other being
> immediately informed, as easily as if he or she had a giant
> screen inside, at the time of the talkies, with remote
> control [*télécommande*] for changing channels and fiddling
> with the colors, the speech dubbed with large letters in
> order to avoid any misunderstanding. For foreigners and
> deaf-mutes. This puerile belief on my part, of a part in
> me, can only refer to this ground—OK, the unconscious,
> if you like—from which there arose objectivist certainty,

this (provisional) system of science, the discourse linked
to a state of science that has made us keep telepathy at
bay [*tenir en respect le télépathie*]. (Te, 2367; Tf, 247)

Another essential distinction within the region of the occult is
the difference between communicating with the dead (orthodox
spiritualism) and receiving telepathic communications from those
who are still alive though at a distance (Freud's examples). This
distinction too is often blurred in practice. Often Freud's examples
are, if not a communication from the dead, a message that tells the
one who receives it that some family member has died. Death is
never far away either in telepathic experiences or in spiritualism.
Martin McQuillan, in a fine essay, "Tele-Techno-Theology,"
observes that Freud's examples of possibly telepathic dreams are
often about not just any deaths, but about soldiers' deaths in World
War I. Freud's telepathy essays were written at the time of collec-
tive post-war trauma after World War I and in the gathering
shadows of World War II and the Holocaust. The latter were to
make Freud an exile, a Wandering Jew, at the end of his life. Freud,
after all, was perhaps the first doctor to analyze in detail the specific
nervous illness brought on by the experience of mechanized war
with its mass killings. It was then called "shell shock." Tens of
thousands, perhaps hundreds of thousands, of our soldiers in Iraq
and Afghanistan are returning with what we now call post-trau-
matic stress syndrome. Now back home, they are killing others and
themselves in record numbers, as many media stories have
reported. More United States soldiers committed suicide during
January 2009, than were killed in Iraq and Afghanistan
combined.[25]

Death is a central motif already in Browning's telepathy poem.
Sludge, at the poem's end, threatens to tell all the world (falsely)
that Horsefall has strangled or poisoned his mother:

. . . ay, you gave her throat the twist,
Or else you poisoned her! Confound the cuss!

Where was my head? I ought to have prophesied
He'll die in a year and join her: that's the way.

I don't know where my head is . . . (CPW, 412)

Mother–son relations turn up again and again in these telepathy
stories, as do heads. An example of the latter is a story Freud tells
in "Psycho-Analysis and Telepathy." This story is important for
Derrida in "Telepathy." It is the legend about Saint Denis who,
decapitated, picked up his head and walked quite a distance with
his head under his arm. Freud's adjective for the picked-up head is
aufgehoben, which delights Derrida because of the Hegelian associ-
ations. *Dans des cas pareils, ce n'est que le premier pas qui coûte*, says
Freud in French, echoing what the custodian of the Saint-Denis
basilica tells visitors. "In such cases as these, only the first step is
costly."

According to my *Petit Robert* French dictionary, this is a prover-
bial expression in French. The following (and last) sentence of the
essay returns to German: *Das Weitere findet sich*. ("The rest sorts
itself out.") I suppose Freud meant that if you can have the least
soupçon of belief in telepathy, or experience of it, you can easily go
all the way, with your head under your arm. Earlier in the paragraph, Derrida says that after Freud has had "the cheek [*le culot*] to
say that his life has been very poor in terms of occult experiences,
he adds: but what a step beyond [*pas au-delà*] it would be if . . .
[*welch folgenschwerer Schritt über* . . .]" (Te, 243; Tf, 253). Then
follows the story about Saint Denis. In for a penny in for a pound,
when it comes to dabbling in telepathy. It is like losing your head
and then walking while carrying it under your arm. The first step
is the hardest. It might be best not to fool around with telepathy
at all, since you can neither believe it nor disbelieve it, as is the case
with ghosts. You say you disbelieve in either at your peril. Saying
you are a skeptic, as Browning does and as Freud sometimes does
(for example in the last sentence of "Dreams and Telepathy": "I
have no opinion on the matter and know nothing about it" [SE,

18: 220]), is the quickest way to raise a ghost or to get a telepathic message.

For diplomatic reasons having to do with his desire to preserve psychoanalysis's purity, Freud more than once suggests a troubled distinction between "thought transference" and telepathy proper. The analyst practices or experiences all the time a species of thought transference, as the patient's unconscious communicates with the analyst's unconscious, while telepathy proper seemed to Freud a "foreign body" within psychoanalysis. It was something in which he both believed and did not believe. He made great efforts to make telepathy a subordinate feature of the interpretation of dreams. It was a question of which was to be master, as Humpty-Dumpty said.[26] Freud felt a lot was at stake in making sure that psychoanalysis was the master of telepathy. He wanted to be sure to be able to incorporate it, even if as a foreign body, into his theory of dreams and the unconscious.

A full reading of what Derrida had to say about the effects of media on culture would take a lot of pages, since the topic comes up repeatedly, not only in "Telepathy," but in *The Post Card*, *Archive Fever*, *Echographies of Television*, *Rogues*, *Philosophy in a Time of Terror*, *Paper Machine*, and so on. The topic obsessed him, even in early works like *Of Grammatology*. I limit myself to the identification of essential features of Derrida's media theory:

Derrida believed that a change in the dominant communications media (for example, from writing and print media to electronic media) changes more or less everything in personal, communal, cultural, and political life. "Envois" is the essential and most immediate context for "Telepathy," since the latter is a misplaced part of the former. A passage in the former claims that "an entire epoch of so-called literature, if not all of it, cannot survive a certain technological regime of telecommunications (in this respect [*à cet égard*] the political regime is secondary). Neither can philosophy, or psychoanalysis. Or love letters" (PCe, 197; PCf, 212). So much for Heidegger, for Freud, and for Derrida's own sending and receiving of love letters in "Envois," or for his writing in "Envois" a postmodern novel in letters and postcards. All three were among the last authors of vanishing genres.

I suppose that when Derrida said, "in this respect the political regime is secondary," he meant that, for example, both terrorists and the endless War on Terror are essentially, not just superficially, determined by the new digital age, the age of cellphones that can be used to trigger bombs, the age of unmanned drones, stuffed with digital devices, and of "smart missiles." One might at first think to challenge this by remembering a famous passage about spiritualism and table-turning, as figures for commodity fetishism, in chapter one, section four, of the first part of Marx's *Capital*:

> It is as clear as noon-day, that man, by his industry,
> changes the forms of the materials furnished by Nature,
> in such a way as to make them useful to him. The form of
> wood, for instance, is altered, by making a table out of it.
> Yet, for all that, the table continues to be that common,
> every-day thing, wood. But, so soon as it steps forth as a
> commodity, it is changed into something transcendent. It
> not only stands with its feet on the ground, but, in
> relation to all other commodities, it stands on its head,
> and evolves out of its wooden brain grotesque ideas, far
> more wonderful than "table-turning" ever was.

I cite this from the Internet, by the way.[27] No need any longer to find it in my printed copy of *Capital*. Marx suggests that spiritualism is a feature of a certain stage in industrialized capitalism and commodity fetishism. Spiritualism was much in vogue just when Marx was writing *Capital*, laboring away in the old British Museum Reading Room. Browning's "Mr. Sludge, 'The Medium,'" attests to that vogue. For us today, Marx asserts, tables can stand both on their legs, and on their heads, in a materialization of the catachrestic prosopopoeias in such phrases as "table leg" and "head of the table." A catachresis is a transferred name for something that has no proper name. The leg of a table is not really a leg, but no other less metaphorical name for those pieces of wood a table stands on exists. Since a "leg" is, literally, the name of the limb of a human being or animal, "leg of a table" is an implied personification of the table. *Prosopopoeia* is the technical Greek name for the figure of speech we call "personification." *Prosopopoeia* means, literally, the ascription of a name, a face, or a voice to something, such as a table, that does not really have these. When tables stand on their heads, they conceive grotesque ideas, presumably Marx's name here for the ideological mystifications of commodity fetishism. These ideas, says Marx, are far more wonderful and "transcendent" than what spiritualist mediums were claiming to make or to allow tables to do, under the guidance of the spirits.

Marx was, I believe, right in what he said about commodity fetishism. The whole gigantic airy Ponzi scheme of derivatives and credit default swaps of subprime mortgages has been built on the shaky material foundation, like a rickety table, of all those homes that were overvalued during the housing boom. Finance capitalism and its underlying commodity fetishism trumps spiritualism. It does so by being better at doing, with the aid of computers these days, what spiritualism claims to do, that is, turn matter into spirit, just as Freud was, in the twentieth century, to declare in a letter of March 1925 to Ernest Jones (cited Te, 260; Tf, 270) that he was an excellent medium. Anything you can do, I can do better. Freud is proud of his gifts as a medium, even though in his telepathy essays he claimed repeatedly that psychoanalysis could explain telepathy scientifically, and spiritualism too. Freud could account for both in a hard-headed, scientific, psychoanalytic way.

What Derrida claims about new telecommunications media, in the "Envois," may, after all, not be too different from what Marx implies. Marx holds that the reigning techniques of production and distribution implacably determine ideology. Derrida's twist on that is to include communications technologies among those ideology-determining factors in a given stage of capitalist technological development. New telecommunications devices have certainly been essential, as I have said, in making possible the current banking crisis. A given tele-technological regime, according to Derrida, outweighs any more or less superficial features of a given political "regime." The new telecom regime triumphing today will make love letters and what we call literature no longer possible. Marx might well have said the same. Capitalism is capitalism, whatever the political regime, just as a culture dominated by email messages and computer programs is the same under any political regime.

Derrida, in both "Envois" and *Archive Fever*, asserts that psychoanalysis would have been impossible except at a time when Freud and all his associates wrote letters to one another and sent them by post. Psychoanalysis was both facilitated and fundamentally limited by a certain stage of technical reproducibility and transmission. I cite the "Envois":

> The end of a postal epoch is doubtless also the end of
> literature. What seems more probable to me [than that
> Poe could have adapted "The Purloined Letter" to an
> informatics age] is that in its actual state psychoanalysis,
> itself, cannot read "The Purloined Letter," can only have
> itself or let itself be read by it [*seulement par elle se faire ou
> laisser lire*], which is also very important for the progress
> of this institution. In any case, the past and present of the
> said institution are unthinkable outside a certain postal
> technology, as are the public or private, that is secret,
> correspondences which have marked its stages and crises,
> supposing a very determined type of postal rationality, of
> relations between the State monopoly and the secret of
> private messages, as of their unconscious effects. That the
> part of "private" mail tends toward zero does not only
> diminish the chances of the great correspondences (the
> last ones, those of Freud, of Kafka), it also transforms the
> entire field of analytic exertion [*l'exercice analytique*]—and
> in both the long and the short term, with all the
> imaginable and unimaginable consequences for the

"analytic situation," the "session" [séance], and the forms of transference. The procedures of "routing" and of distribution, the paths of transmission, concern the very support of the messages sufficiently not to be without effect on the content, and I am not only speaking of the signified content. (PCe, 104; PCf, 114–15)

Many readers will doubtless resist what Derrida so hyperbolically says here. Can what we call "literature" really depend on the technology of the postal system, the political institutions that make the modern postal system possible, and the secrecy that system supposedly secures? Surely literature still flourishes today and will continue to do so in the age of the Internet. No, says Derrida, the shift from snailmail letters to email, that is, a radical shift in media, spells, sooner or later, the end of the institution, literature, that was historically intertwined with the epoch of handwritten letters sent by the post. The last sentence of the citation presumably refers to Derrida's notion of the basic reason for this: media are performative, not simply constative. Their specific matrices make something happen. The "content" of whatever is said or graphically displayed in a given medium at a given time and in a given context is affected by the medium used, that is, by the "very support of the messages." A given medium is not the passive carrier of information. A medium actively changes what can be said and done by its means.

Derrida's version in "Telepathy" of his hypothesis that a change in media changes everything else is to claim the great significance of the fact, asserted more than once, that all the evidence Freud proffers to support his quasi belief in telepathy involves "the constant association, at least in terms of the figures, comparisons, analogies, and so on, between a certain structure of telecommunications, of the postal technology (telegrams, letters and postcards, telephone) and the material that is today situated at my disposal when I hear talk about telepathy" (Te, 252; Tf, 262). "My" refers to Freud himself. Derrida here speaks for Freud, as Freud, as a medium through whom the dead Freud speaks. I shall return to the significance of this ventriloquism later, in the third part of this book. For Derrida, both Freudian psychoanalysis and telepathy were determined by a certain stage of telecommunications: the epoch of the postal system and the telephone. "One can only dream or speculate," says Derrida in *Archive Fever*, "about the geo-technological shocks which would have made the landscape of the psychoanalytic archive unrecognizable for the past century if, to limit myself to these indications, Freud, his contemporaries, collaborators and immediate disciples, instead of writing thousands of letters by hand, had had access to MCI or AT&T telephonic credit cards, portable tape recorders, computers, printers, faxes, televisions, teleconferences, and above all E-mail."[28]

Yes, one can only dream, speculate—or google it.

On the next pages of *Archive Fever*, echoing or being echoed by other such assertions he elsewhere makes, Derrida expands what he says about psychoanalysis's confinement within a certain communication technology. He affirms, extravagantly, though not without justification, in the event, that email is changing everything in human personal, communal, and political life. Email is causing "radical and interminable turbulences" (AF, 18; MA, 35–6):

> [E]lectronic mail today, even more than the fax, is on the way to transforming the entire public and private space of humanity, and first of all the limit between the private, the secret (private or public), and the public or the phenomenal. It is not only a technique, in the ordinary and limited sense of the term: at an unprecedented rhythm, in quasi-instantaneous fashion, this instrumental possibility of production, of printing, of conservation, and of destruction of the archive must inevitably be accompanied by juridical and thus political transformations. These affect nothing less than property rights, publishing and reproduction rights. (AF, 17; MA, 35)

Derrida, it is worth noting, though he wrote on the computer for the last fifteen years of his life, strongly resisted email's transformations of humanity. He resisted it as instinctively and forcefully as Heidegger resisted the typewriter. Though Derrida had a somewhat unreliable fax machine, he never, ever, used email, to my knowledge. He preferred writing letters and comments on student papers with pen in hand, though, somewhat reluctantly, toward the end of his life, he typed some of them because he recognized that his handwriting was nearly illegible. Paul de Man and I, in those far distant days at Yale, in the '70s and '80s, used to spend a whole lunch together trying to decipher some two-page letter or other Derrida had sent me. The small army working today to decipher the Derrida archive, for example his early handwritten seminars, have their work cut out for them.

Derrida believed, as the passages just cited attest, that a given medium is not just a neutral recorder and transmitter of information. A medium decisively determines what can be said by way of it. That means that the medium is not just a purveyor of constative facts. Something embodied in the *teknē* of a given medium is performative. It makes something happen. The medium is the maker. It is as though, for example, Mr. Sludge's fraudulent reception of a message from the spirit world prophesying that his patron Hiram H. Horsefall will die in a year were actually to bring about his death, as a cat kills a mouse by frightening it to death, or as Freud certainly did his health no good by continually predicting that he would die on such and such a day and then altering the day when he did not die after all. Or rather, the media themselves bring about such effects, whatever the messages they carry. It is not so much that, as Marshall McLuhan said, the medium is the message, as that the medium is the felicitous performative. It does something with words or other signs, rather than just transmitting a message. That, I take it, is what Derrida means in the passage from "Telepathy" I have cited in my epigraph when he says that what mediums of all kinds declare, predict, or say they foresee *"forms part of"* that saying and therefore, "participating in the thing, they also

24

provoke it." The medium itself works performatively to bring about what the prophecies expressed by their means predict. More radically, they do not even need to express this prophesy in constative, comprehensible form.

Archive Fever says this even more forcefully:

> [T]he archive, as printing, writing, prosthesis, or
> hypomnesic technique in general is not only the place for
> stocking [*stockage*] and for conserving an archivable
> content *of the past* which would exist in any case, such as,
> without the archive, one still believes it was or will have
> been. No, the technical structure of the archiving archive
> also determines the structure of the *archivable* content
> even in its very coming into existence [*son surgissement
> même*] and in its relationship to the future. The archive
> produces as much as it records [*enregistre*] the event. This
> is also our political experience of the so-called news media
> [*media dits d'information*]. (AF, 16–17; MA, 34)

The media, manipulated by Bush–Cheney, contributed essentially to the invasion and occupation of Iraq.[29]

This conviction of the performative efficacy of the medium itself, whatever is said in it, is explored in Derrida's attempt to articulate to the "you," the addressee, in the early pages of "Telepathy," a concept of time in which the future event already, telepathically, forms part of the present. This Derridean notion of time is familiar by way of his celebrated neologism, *différance*. Derrida's time is a permutation of Heidegger's time that moves forward into the future in order to come back to the past, as formulated in *Sein und Zeit*. Heidegger's time holds all the horizons of time within one mobile unit. Derrida's time and Heidegger's time are similar, but by no means identical. Heidegger's time is grounded in *Sein*, Being with a capital B. Derrida's time is created out of performative media, the media as makers, including makers of human time. On each occasion when a given medium is used, that use creates its own ground and its temporal *différance*. That ground did not preexist the performative enunciation, whereas Heidegger's *Sein* is, was, and always will be, even though it may, during a given historical epoch, be forgotten or occulted.

The presuppositions about time in "Telepathy" disqualify my suggestions, made earlier in this book, that a given historical state of the media and of its surrounding culture has determined a focus on past, present, or future. For Derrida, all these dimensions of time are versions of a universal apocalyptic time in which the future is already in a sense present in the present and produced performatively by it, in Derrida's version of Heidegger's "ecstasies" of time. Telepathy, he asserts, depends on this doubling of distance and proximity, *tele*pathy and tele*pathy*, *Fort/Da*.

I conclude this section of this book with an attempt to identify just what Derrida's conception of apocalyptic telepathic time is, and just what this implies for transfers and correspondences.

The apocalypse that interests Derrida in "Telepathy" is performatively triggered by a piece of writing that announces it, prophesies it, as a future event:

> It wasn't sufficient to foresee or to predict what would indeed happen one day, forecasting is not enough [The last phrase is in English in the original], it would be necessary to think (what does this mean here, do you know? [*tu sais, toi?*]) what would happen by the very fact of being predicted or foreseen, a sort of beautiful apocalypse telescoped, kaleidoscoped, triggered off at that very moment by the precipitation of the announcement itself, consisting precisely in this announcement, the prophecy returning to itself from the future of its own to come [*à-venir*]. The apocalypse takes place at the moment when I write this, but a present of this type keeps a telepathic or premonitory affinity with itself (it senses itself at a distance and warns itself of itself) that loses me on the way {*me sème en route*} and makes me scared. (Te, 227; Tf, 238–9)

This strange notion of telepathic anticipation is closely articulated to the idea that a letter creates the self both of the writer and of the addressee who takes the letter as addressed to him or her and thereby becomes the person the letter invokes. It also is analogous to the idea, expressed in my epigraph, that a prophetic telepathic "message" is not a comprehensible assertion, but a performative enunciation or annunciation. It predicts something that will happen at a distance and at a later time. At the same time, it is something that happens right now, in the production of the annunciation, for example at the moment when Derrida wrote those words. Telepathy produces, right now, what it predicts.

Derrida, quite properly, connects this idea with a series of proper and common names in Freud's telepathy essays. This series starts with Dr. David Forsyth. Forsyth was an English analyst who left a visiting card with Freud when the latter was in the midst of a *séance*. The sequence continues with Galsworthy's *Forsyte Saga*, and then goes on through the nickname of Freud's patient being analyzed when Dr. Forsyth called on him. It includes some common nouns in English and German that are punned on in the proper names: "Forsyth, . . . Forsyte, foresight, *Vorsicht*, *Voraussicht*, precaution, or prediction [*prévision*], . . . Herr von Vorsicht" (Te, 260, 261; Tf, 269). The latter is the nickname given, by a "virgin," to the patient Freud was seeing when Dr. Forsyth left his card. The girl ironically gave him the name "because of his prudent or discreet [*pudique*] reserve" (Te, 261; Tf, 260).

Are all these "f's" and "v's" and "p's" meaningless coincidences? Perhaps, but Freud (and Derrida) did not think so. Derrida thought any one of these names called up the whole string of them and produced the things they anticipated. This happens by a sort of boomerang telepathy, the prophecy returning to itself from the future of its to come, sometimes with an auto-destructive effect.[30]

Another example of such a string is triggered by the strange dream Derrida reports at the beginning of "Telepathy": "I had a premonition of something nasty [*vicieux*] in it, like a word, or a worm, a piece of worm that would be a piece of word, and that would be seeking to reconstitute itself by slithering [*en rampant*], something tainted [*vicié*] that poisons life. . . . I felt, from a distance and confusedly, that I was searching for a word, perhaps a proper name (for example, Claude, but I do not know why I choose this example right now, I do not remember his presence in my dream)" (Te, 226; Tf, 237–8). *Rampant* is a "vicious," "viscous," and "tainted" word in which two meanings try, not wholly successfully, to "reconstitute" themselves into a single unit. *Rampant* means "slithering," all right, as the translator (Nicholas Royle) says, for example as a worm or snake moves sinuously forward, or as soldiers may crawl forward on their stomachs, but it also means "to rise up," as in heraldry "a lion rampant" designates a lion "rearing on the left hind leg with the forelegs elevated, the right above the left" (*American Heritage Dictionary*), as if prepared to strike or to climb. The word does what Derrida is talking about, that is, express two meanings trying to come together, just as "Claude" is both a female name and a male name. "Claude" is probably a reference to Derrida's male and female cousins of that name, explained as a clue to Derrida's "impossible homosexuality" in section 31 of *Circumfession*.[31] But who knows? Maybe he was thinking of some other Claude (such as his friend Jean-Claude Lebensztejn, a distinguished art historian and critic). Maybe Derrida knew some other Claude, or just happened to dream of the name, though it is unlikely that the name was completely unmotivated.

Much could be said about this dream, including details I have omitted about Derrida's hair falling out, or "your hair," and being unsuccessfully knotted to keep it in, but I refrain from that in order to suggest another reason for the name "Claude." That reason emerges a few pages later in a series like the "Forsyth" one: "*glas* . . . (*cla, cl, clos, lacs, le lacs, le piége, le lacet, le lais, là, da, fort, hum* . . . [cla, cl, closed, lakes, snare, trap, lace, the silt, there, here, yes, away, hmm . . .])" (Te, 234, trans. modified; Tf, 245), and, finally, on the next page, *claudication* (Te, 235; Tf, 246). That word means "limp," as in Derrida's limp after his fall from his son's skateboard and as in the limp at the end of Freud's *Beyond the Pleasure Principle*, to which I shall return. *Caler*, which means "stall," as in Derrida's repeated *Je cale* ("I'm stalled") (Te, 230; Tf, 242), is an important item in this series. The primary elements in the sequence come from *Glas*, Derrida's big book about Hegel and Genet. *Glas* means "funeral knell," but also, like other words in the series, it mimics the choking sound you make when you are drowning or being strangled. The name "Claude," Derrida seems to be suggesting, if the pieces of the worm in his dream had come together to make this word, which luckily they did not, would have triggered, like a cascade or waterfall, all the other words in the series. Therefore the assembled word "Claude" would have worked performatively to bring about Derrida's drowning or strangling, as in his adolescent fantasy of drowning himself in the *glu*, or slime, of a pond: "I forgot. The first verse I published: '*glu de l'étang lait de ma mort noyée.*' ('glue of the pool milk of my drowned death')."[32] This would have been the boomerang effect with a vengeance, Derrida's dreamed word turning back on him, if he is so unlucky as to remember it, the annunciation returning on the announcer to choke or drown him, unmaking the maker.

"Mr. Sludge, *c'est moi*"

I CAN still recall vividly how Freud said to me, "My dear Jung, promise me never to abandon the sexual theory. This is the most essential thing of all. You see, we must make a dogma of it, an unshakable bulwark." He said that to me with great emotion . . . In some astonishment I asked him, "A bulwark—against what?" To which he replied "Against the black tide of mud"—and here he hesitated for a moment, then added—"of occultism." (C. G. Jung, *Memories, Dreams, Reflections*)[33]

THIS section tests out Derrida's proposition about the socio-politico-personal-performative effects of a change in media by looking at an example of the collision of two media: Robert Browning's "Mr. Sludge, 'The Medium.'" Had Browning read Derrida, as I shall show, he might have called his poem, "Mr. Glu, 'The Medium.'" The third section of this book will focus on at the strange rhetoric of Derrida's "Telepathy," its exploitation of the written letter, in the sense of "epistle," as a specific medium, as opposed to the overt themes those rhetorical devices are used to "express." It is perhaps in what can be broadly called the "conventional rhetoric" of a given medium, including visual ones like cinema, that the each medium's power as a "maker" is embodied. This is true whatever those rhetorical devices are used to express in a given case.

There would be much to say about Browning's poem. An extended reading would be profitable, worth doing. It would repay the effort. "Mr. Sludge, 'The Medium,'" is one of Browning's most brilliant dramatic monologues. It was written at a highpoint of the mediumistic *séances'* Victorian vogue. I focus rather narrowly, however, on the way the poem is a performative, a curse, a "flyting," as they used to be called, an attempt to put the double whammy on Daniel Dunglass Home. Home was a highly successful spiritualist medium of the mid-nineteenth-century. Browning casts his spell on Home by exposing him in the guise of a fictive, ghostly, counterpart, Mr. Sludge. Sludge is the spokesperson in the monologue.

I approach Browning's poem by way of a question: Why did Browning hate Home so much? He hated Home enough to personify him as Sludge, that is, as a form of *glu*. Sludge is perhaps the most detestable of what have been called, by Browning specialists, the poet's "casuists." These casuists are speakers in his monologues who adduce plausible but specious and hateful forms of self-defense. The name recalls those old Jesuitical casuists Browning, as a Protestant, would have found so sinister, in their ability to find excuses for almost any unrighteous act. Browning hated Home enough to call him a "dung-ball." This was no doubt a play on Home's middle name, "Dunglass." This epithet appears in a letter of 1863: dung, sludge, excrement, the opposite of disembodied spiritual purity. "Mr. Sludge, 'The Medium,'" is an elaborate way to call Home, to say it in the vernacular, "a shit." Browning threatened in another letter (of 1871) to kick Home if he ever encountered him in the street, though he feared doing that might dirty his shoe: "If I ever cross the fellow's path I shall probably be silly enough to soil my shoe by kicking him."[34]

Why this extravagant and seemingly excessive hatred? Browning was present in 1855 with his wife, a strong believer in spiritualism, at one of Home's *séances*. This occurred in the drawing room of the Browning's friends, the Rymers, in Ealing (London). At that original *séance*, a clematis wreath, that was resting on the table around which they all sat, rose in the air, moved across to where Elizabeth Barrett Browning sat, and gently settled itself on her head. Browning, when he next encountered Home, in his own drawing room, a few days later, pointedly refused to shake hands with him.

D. D. Home was born in Edinburgh. He moved to the United States at the age of nine. He was brought up in Troy, New York, and then in Norwich, Connecticut. His mother had second sight, and her son inherited that gift. He began his "mediumship" at the age of eighteen and continued it the rest of his life. After many successful *séances* in the Northeast of the United States in the 1850s (Springfield, Boston), Home moved to Europe, where he had a spectacular career during the 1850s and 1860s. He gave *séances* in England, in Florence, and in Paris, including one for "their Majesties at the Tuileries, where manifestations of an extraordinary nature occurred." He was received into the Roman Catholic Church by the Pope himself. He met the Queen of Holland. He married in succession two Russian noblewomen. He held a *séance* for the Czar while he was at St. Petersburg. Home published two volumes of *Incidents in My Life* (1863, 1872). The second of these gives his version of his encounters with Robert Browning. Though often accused of being a fraud, he was never found out. [35] Home was depicted in Victorian satiric drawings as levitating tables and people. Here is one example:

Levitating a Table, with Client [36]

Home must have been as gifted as any skilled magician at sleight of hand and at the ability to distract the audience's attention. Freud identifies these two skills as requisites in successful mediums, magicians—and psychoanalysts—when they are moving wreaths through the air, or pulling rabbits out of hats, or pigeons from their sleeves, or repressed memories from a patient's unconscious. Freud was himself no mean hypnotist.

I mention some details of Home's career to confirm that he was more than nominally successful and to stress the fact that he was never exposed as the imposter Browning firmly believed him to be.

"[H]e," Browning wrote of himself in a letter, "is hardly able to account for the fact that there can be another opinion than his own on the matter—that being that the whole display of 'hands,' 'spirit utterances,' etc. , were a cheat and imposture" (cited in DeVane, 310). Well, all right, but why carry his distaste to such lengths?

The obvious answer is that Browning was jealous of Home's power over his wife. It may be that he feared Home was alienating his wife's affections. It may even be that, as Home maliciously and probably falsely suggests, Browning was jealous of his wife because he had wanted the spirits to put the wreath on his own head. He should be the crowned poet, not she. No doubt all these factors may have been part of the source of his rage. Nevertheless, they hardly explain or justify Browning's violent anger against Sludge the dung-ball. The poem Browning wrote about Home suggests a different answer to my question.

"Mr. Sludge, 'The Medium,'" was written around 1859–60, before Elizabeth Barrett's death in 1861. It was published in 1864, in *Dramatis Personae*, after her death. We have no evidence that she ever saw the poem. It might well have caused her pain, though her belief in spiritualism had apparently weakened by the time of her death. This was "brought about," says her son, "in a great measure, by the discovery that she had been duped by a friend in whom she had blind faith" (de Vane, 312). Could that "friend" have been Home, or Home's sponsor, Mrs. Rymer?

Browning's poem is a characteristically dense and complex example of the poet's longer casuistical dramatic monologues. It is 1525 lines long, in blank verse. The poem takes its place in the context of Browning's other dramatic monologues, in their swarming multitude. In these poems, Browning "speaks for," or "speaks in the name of," an immense number of historical or imaginary persons, from Aristophanes in one poem ("Aristophanes' Apology"), to a witness of the death of St. John in another ("A Death in the Desert"), through those Renaissance painters Browning learned about from Vasari's *Lives of the Painters*, Fra Lippo Lippi and Andrea del Sarto, to the speakers in the collection of twelve long monologues by the participants in Browning's Roman murder story, *The Ring and the Book*, to Don Juan, in "Fifine at the Fair," to Napoleon the Third, in "Prince Hohenstiel-Schwangau, Saviour of Society," to—Daniel Dunglass Home in "Mr. Sludge."

Browning, in the preface to *Dramatic Lyrics* (1863), says such poems are "though often Lyric in expression, always Dramatic in principle, and so many utterances of so many imaginary persons, not mine" (CPW, 163). Aristophanes is hardly an "imaginary person," but the speech Browning ascribes to him, as to his other monologuists, *is* imaginary. It is a product of the sympathetic imagination, or of Keatsian negative capability, that is, the power the early twentieth-century French critic Charles du Bos, in a fine phrase, called *l'introspection de l'autrui* ("the introspection of the other person").[37] Browning inherited the idea of the sympathic imagination from the Romantic poets, and he arrogated that power to himself in his first significant poem, "Pauline" (1833). There his persona boasts, "I have gone in thought/Through all conjuncture. I have lived all life/When it is most alive, where strangest fate/New-shapes it past surmise—the throes of men/Bit by some curse or in the grasps of doom/Half-visible and still-increasing round,/Or crowning their wide being's general aim" (CPW, 8). Though this is a little hectic, even for a twenty year-old novice poet much influenced by Shelley, it already contains the program for Browning's abundant poetic production in the decades that followed. That production is based on the poet's claim that he is able to enter into the consciousness and feelings of some other person, imaginary or real, alive now or long dead, at some crucial moment in his or her life. The poet can then speak for that other. He can become a medium through which the other speaks.

A dramatic monologue is called "dramatic" because it takes its speaker at a moment when he or she is instigated to speak by being in some shrewd situation or other. In "Mr. Sludge, 'The Medium,'" Browning imagines, quite fictitiously, that Home/Sludge has been found out cheating by his Boston patron, Hiram H. Horsefall.

34

Sludge has to do some fast talking to keep Horsefall from exposing him. He also needs to get Horsefall to give him enough money to take the boat to England so he can restart his fraudulent career there afresh. Sludge puts forth every excuse he can think of: that this is the first time he has cheated; that his patrons, when he was too young to know better, encouraged him to claim he had seen supernatural phenomena when he had not; that he was tempted by the luxury his career as a medium has afforded him; that he has only catered to the boundless credulity in people generally about access to the spirit world; that he is not sure that amidst all his lies there may be some grain of truth, some actual influx from the spirit world mixed with all the fictions and spurious raps, table turnings, spirit voices, and spirit writings; that belief in signs, omens, miracles, ghosts, and interventions from "beyond the veil" is an integral part of Christianity, so Sludge is only being a pious Christian in his mediumhood, and so on, in a dazzling display of self-defensive rhetoric.

Much of what Sludge says anticipates in detail the arguments Freud makes in expressing his contradictory, convoluted, and embarrassed yes/no relation to telepathy. That expression takes the form of some letters and of his four essays or what Derrida calls "fake lectures" on telepathy. They are "fake" because Freud only presented one of them orally, though all are written like lectures. Sludge's rhetoric, like Freud's, is a kind of contradictory overkill. The various forms of apology Sludge offers are not mutually compatible. Moreover, in the course of his speech Sludge gives away his procedures, for example the tricks whereby he picks up little bits of information about all the people he meets. He can then produce these bits as evidence of thought transference, proof of his ability to know things he "could not" know (CPW, 402). Freud was to do the same. One might argue that the basic technique of psychoanalysis depends on its skill in giving meaning to apparently trivial details in a dream or in everyday behavior, slips of the tongue, for example, or modes of metonymic displacement and metaphorical condensation. The consonance between what

Browning has Sludge say and what Freud says in his telepathy essays is little short of amazing. It is not clear, however, whether this confirms Browning's genius, or Home's, or Freud's, or marks all three as impostors.

Among these conflicting and self-canceling, casuistic, arguments is the claim by Sludge that he is, after all, only doing what novelists and poets do and are universally acclaimed for doing. Sludge, he boasts, is even doing it better than the poets do because he actually becomes the fictitious or historical characters for whom he speaks. This comparison with the lies poet-makers tell recurs throughout Sludge's speech. What he has done, says Sludge, is "not so very false, as falsehood goes,/The spinning out and drawing fine, you know,—/Really mere novel-writing of a sort,/Acting, or improvising, make-believe,/Surely not downright cheatery" (CPW, 402). "When you buy/The actor's talent, do you dare propose/For his soul beside? Whereas, my soul you buy!/Sludge acts Macbeth, obliged to be Macbeth,/Or you'll not hear his first word!" (CPW, 404). Sludge imagines, speaking by way of the metaphor of painting, novelists making much money and reaping much praise by inventing extravagant narratives about Sludge himself: "That man would see the whole world roll/I' the slime o' the slough, so he might touch the tip/Of his brush with what I call the best of browns—/Tint ghost-tales, spirit-stories, past the power/Of the outworn umber and bistre!" (CPW, 405)

"Best of browns!" Excrement, as in "dung-ball," or "sludge," is never far away in Browning's imagination of "Mr. Sludge, 'The Medium'": "He [the muck-raking novelist]'s the man for muck;/Shovel it forth, full-splash, he'll smooth your brown/Into artistic richness, never fear!" (CPW, 405) An unfortunate, and no doubt inadvertent or "unconscious," overtone in the echo of "Browning" and "the best of browns" may make the reader uneasy, if he or she happens to notice it. In Sludge's figure, ascribed to him by Browning, excrement becomes the paint with which a despicable novelist or poet, "the other [along with the philosopher] picker-out of pearl/From dungheaps,—ay, your literary man"

(CPW, 405), embellishes and embrowns his narrative. Just who is the dung-ball now?

A little later in the poem, a final example of the recurrent comparison of mediums to poets appears. It is a comparison that winds through the poem, like a trail of sludge. Sludge says, "so, Sludge lies!/Why, he's at worst your poet who sings how Greeks/That never were, in Troy which never was,/Did this or the other impossible great thing!/He's Lowell—it's a world (you smile applause)/Of his own invention—wondrous Longfellow,/ Surprising Hawthorne! Sludge does more than they,/And acts the books they write: the more his praise!" (CPW, 411)

"Mr. Sludge, 'The Medium,'" exposes what most angered and frightened Browning about Sludge's mediumship. Home, the poem tells us, was doing just what Browning was doing, for example in the poem we are at that moment reading. Browning, like Sludge, serves as mouthpiece or medium through which the other speaks, in a spooky raising of ghosts: Aristophanes, Fra Lippo Lippi, Guido Fransceschini, and all the rest, including Sludge. The difference, Browning seems to have feared, is that Home was doing it better, in a different way, with a different teletechnology, and having a wider influence, even over Browning's beloved Elizabeth Barrett, not to speak of getting better paid for it.

As Sarah Kofman has recognized,[38] Freud saw telepathy, along with other forms of the occult, as a challenge to psychoanalysis that must be encompassed and reduced, subjected to science. Telepathy must be turned into one subordinate tool in the analyst's toolbox. Freud seems to have wondered about who is the better medium, the psychoanalyst or the spiritualist. He claimed, in a letter to Ernest Jones already cited, to have discovered experimentally that he was a gifted medium (cited Te, 260; Tf, 270). Just as the occult challenged psychoanalysis, so spiritualism was a challenge to Browning's poetry. Home, Browning, and Freud made the dead speak, by negative capability, as opposed to the Wordsworthian egotistical sublime, in which the poet speaks only for himself, as a dominating "I."

Browning's hatred of Home was evidence of competing tele-communication-technologies ferociously scrambling for what Freud claims, in one of the *Introductory Lectures*, all men (sic!) want and get through art: power, honor, and the love of women. Having in their works presented fictitious stories about acquiring these and thereby having given their readers the pleasure of imaginary possession of them, male artists then reach through the admiration of their readers what they at first had only in imagination. The artist "makes it possible for other people once more to derive consolation and alleviation from their own sources of pleasure in their unconscious which have become inaccessible to them; he earns their gratitude and admiration and he has thus achieved *through* his phantasy what originally he had achieved only *in* his phantasy— honor, power, and the love of women" (SE 16: 376–7).

Sludge, Browning fears, employs more successfully than he a better technology for doing this. Therefore Browning is excessively angry, just as Freud fears telepathy might put psychoanalysis out of business. In a similar way, Derrida foresees, in "Telepathy" and in the "Envois" part of *The Postcard*, that new teletechnologies will put an end to what he is at that moment doing, what it is his vocation to do, that is, write literature, love letters, and philosophy. In a contemporary echo of this conflict of media, the recording industry, these days, along with the printed books industry, is threatened with extinction by iPods, iPhones, Blackberries, MP3 players, e-books, and freely available online files of an amazing array of printed texts.

Browning inadvertently gives himself away in the act of providing Sludge rope to hang himself, that is, by giving him so many eloquent rationalizations for his impostures. Browning is in effect confessing, "Mr. Sludge, *c'est moi.* We are both gifted embrowners, for example in an overtone I have slyly inserted in the name, Horsefall. Horsefall, to speak colloquially, is what falls from a horse: dung-balls or horseshit. I do not need to spell out the connotations of 'Sludge.' My poem not only slanders Home; it also is self-slander."

No reader of "Mr. Sludge" who knows Browning's other poetry could fail to identify this as a poem by Robert Browning. As other readers have noticed, Browning ascribes to Sludge his own gift for powerful rhetoric supported by grotesque metaphors. One example out of many is the double figure Sludge uses to describe his ability to pick up apparently insignificant details about people he meets and then to use them in *séances* to convince clients of his supernatural powers. Sludge compares this first to the secret occupation of a cobbler in Rome who worked outside his shop in the most innocent-looking way possible. That lowly shoemaker was really a police spy. He made a living informing on people who inadvertently gave themselves away by something they said or by some self-revealing gesture as they passed his shop. That figure is then figured in a figure that only Browning is likely to have invented. It bears his trademark: "His [the cobbler's] trade was, throwing thus/His sense out, like an ant-eater's long tongue,/Soft, innocent, warm, moist, impassible,/And when 't was crusted o'er with creatures—slick,/Their juice enriched his palate" (CPW, 403).[39] This is just what both Sludge and Browning do, each in his own way, and with different communication technologies.

Another figure from the natural world that has Browning's personal stamp is used by Sludge to describe the way he has intimations of the spirit world. The figure, in ascribing telepathic power to cranes, anticipates the one about the "great insect communities" that Freud uses in the passage naturalizing telepathy I have cited above: "I guess what's going on outside the veil,/Just as a prisoned crane feels pairing-time/In the islands where his kind are, so must fall/To capering by himself some shiny night,/As if your back-yard were a plot of spice—/Thus am I 'ware o' the spirit-world" (CPW, 409). Sludge speaks both for his own procedure and for Browning's when he says that in a *séance* he often begins speaking in the voice of the ghost that putatively speaks through him and then reverts to his own voice, though nobody much notices the difference: "I've made a spirit squeak/In sham voice for a minute, then outbroke/Bold in my own, defying the imbeciles—

/Have copied some ghost's pothooks, half a page,/Then ended with my own scrawl undisguised" (CPW, 403). Browning, no adept reader can fail to note, makes Sludge speak Browningese.

Browning most betrays his kinship with Sludge (the way he speaks for himself in speaking for Sludge), against the whole apparent polemical purpose of the poem, by ascribing to Sludge the arguments for spiritualism that would have carried most weight with Browning himself. The argument that there might be a little something to spiritualism, after all, in spite of its fraudulent side; the argument that after all it only follows Biblical beliefs about miracles, signs, omens, and spectral appearances; Sludge's use of the Biblical doctrine that there is a Providence in the fall of a sparrow—all this Browning would have found persuasive, or almost persuasive. This is so in spite of his adherence to what the Bridgewater Treatises, mentioned by Sludge, had to say about the disappearance of miracles in these latter days and about the providential fitting of the natural universe to the needs of man, without any necessity for anything "unnatural" or uncanny.[40]

It almost seems as if Browning in the course of his poem makes himself a convert to spiritualism, or at least makes himself someone who, like Freud after him, must say that he neither believes nor disbelieves. "I have no opinion on the matter and know nothing about it," says Freud at the end of "Dreams and Telepathy" (SE 18: 220), in a declaration already cited. This is a *Verneinung*, or denegation, if there ever was one. Browning, like Freud, talks himself into the corner of needing to believe in something like spiritualism and telepathy. He needs to believe, that is, if he is going to sustain his vocation as a poet-medium who can "throw his sense out," and through whom the dead can speak. He has to be able to claim that he is a better medium than Home.

Browning's argument, via Sludge, for spiritualism is also just the argument that Freud makes in his own way. Sludge's superstition is like Freud's superstition and like his reasons for believing in telepathy. The evidence is that Freud really did believe, somewhat shamefacedly and anxiously, and in spite of all his disclaimers. Freud, Sludge, and Browning also all have similar contradictory explanations for how clairvoyants, poets, and psychoanalyists can know things about other people they apparently could not possibly know, even though the mediums or media for doing this are different in each case. Derrida's mediumhood is a little different, though no less contradictory. To that I now turn.

Derrida as Medium

. . . telepathy is the interruption of the psychoanalysis of psychoanalysis. (Te, 260; Tf, 270)

So psychoanalysis (and you're still following the fold line [the line between external data about telepathy in Freud's essays and the way these essays are "autobio-thanatography" (JHM)]) resembles an adventure of modern rationality set on swallowing *and* simultaneously rejecting the foreign body named Telepathy, assimilating it and vomiting it up without being able to make up its mind to do one or the other. (Te, 261; Tf, 270)

Fort: Da, *tele*pathy against tele*pathy*, distance against menacing immediacy, but also the opposite, feeling [*le sentiment*] (always close to oneself, it is thought), against the suffering of distancing that would also be called telepathy. (Te, 259; Tf, 268)

D ERRIDA'S "Telepathy" is an exceedingly strange document. It is more than a little wild. Even its genre is hard to identify. Is it a philosophical essay? A heavily edited collection of actual letters and postcards, weird love-letters that one or another of the several Derridas may have sent to someone or to several someones? A literary-critical reading of Freud's essays on telepathy? Two fictitious dramatic monologues embedded one within the other? Or is it best thought of as, from one end to the other, an apostrophic performative that invokes, by manifold rhetorical devices, the complicit reader, the *toi*, it needs? Perhaps "Telepathy" is all these things at once, in a more or less impossible combinatory possibility, a radical "perhaps." A full reading of "Telepathy" would be more or less interminable, since it is such a rich and complex text. One way to read it is as Derrida's attempt simultaneously to assimilate and to vomit up both Freudian psychoanalysis and telepathy as foreign bodies within deconstruction, without being able to decide which to do. If Freud could say, in Derrida's phrase, "*moi—la psychanalyse*," Derrida could say, "*moi—la déconstruction*."

"Telepathy," as I have said, was originally supposed to have been part of "Envois," in *The Post Card*. It uses the same vocabulary and ideas as its parent text. Consequently, Alan Bass's admirable "Glossary" of keywords in "Envois" (PC, xiii–xxix) gives some idea of the linguistic density of "Telepathy." These recurrent words are switching points, such as *envois*, *cale*, or *destin*, each with its associated terms. Bass's glossary also indicates why it is often useful to refer to the French original when talking about "Telepathy." An example of this need is the play on the word *voie*, buried in *envois*. This is exuberantly played on in one passage of "Telepathy": "Wait, here I interrupt a moment on the subject of his 'legacies' [*legs*] and of everything I'd told you about the step [*pas*], the way [*voie*], viability, our viaticum, the car and *Weglichkeit* [a reference to all Heidegger has to say about *Weg*, German for "way" or "path" (JHM)], and so on" (Te, 237). Here is this passage in French: "*Attends, là j'interromps un instant au sujet des ses 'legs' et de tout ce que je t'avais raconté sur le pas, la voie, la viabilité, notre viatique, la voiture et la* Weglichkeit, etc." (Tf, 248). Though "legs" is French for "legacies," nevertheless, by a translinguistic pun, it requires legs to take steps on the way, or even to drive a car, *voiture*, literally "way-goer," in French.

42

I shall in this section concentrate on two features of "Telepathy": the way it enacts the theory of apostrophe it enunciates, and the way it contains an intermittent dramatic monologue in which Derrida speaks, mediumistically and telepathically, for Sigmund Freud. Derrida raises Freud from the dead, just as Freud spoke, mediumistically and telepathically, for the quasi-patients who wrote him letters about their telepathic experiences, just as Browning was the medium through which (or through whom) Sludge spoke, and just as I speak here for Derrida and for all those others who speak through him.

Reading "Telepathy" right is a question of understanding correctly its linguistic form, its rhetorical procedures. As we know, the medium is the maker. The medium is the condition of felicitous performativity, if there ever is such a thing. Untangling the rhetorical procedures and their functions in this essay is, however, a challenging task. It is perhaps even in the end an impossible one. Derrida uses the word "scenario" to name what interests him more in Freud's correspondence and in Freud's fake lectures about telepathy than the thematic content of these documents. Speaking of Freud's letters to his friend Wilhelm Fliess, Derrida (or one or another of his spokespersons) says: "The history of this transferential correspondence is unbelievable: I'm not talking about its content, about which there has been plenty of gossip [I suppose Derrida means speculations about the homosexual nature of Freud's intimate friendship with Fliess. (JHM)], but of the scenario—a postal, economic, even banking, military as well, strategic scenario—to which it has given rise and you know that I never separate these things, especially not the post and the bank, and there is always some training [*de la didactique*] in the middle" (Te, 231; Tf, 242). This is a reference to the complicated history of what happened to Freud's letters to Fliess when Marie Bonaparte bought them from the person to whom the malicious Frau Fliess had sold them. Bonaparte refused to sell them back to Freud, who was at that point her training analyst. He had offered her half what she paid for them. She smuggled the letters, under the eyes of the Gestapo, from a bank in Vienna to Paris, and then to London in a waterproof bag, with flotation inserted, over the dangerous English Channel during wartime. Ultimately they ended up in the Library of Congress. Where else? Access to them is limited. Why? What damage could seeing them do anyone, in these latter days? Perhaps it would be embarrassing to Matthew Freud and other Freud descendants now living in England, if the word got around that their great-grandpapa was gay or bisexual, if indeed he was.

Even the provenance of "Telepathy" is strange. It was, Derrida tells the reader in an initial note, intended to be part of "Envois," as I have said earlier, but the "cards and letters" (Te, 423; Tf, 237) that were the basis of the final manuscript, got mysteriously but accidentally mislaid. They then suddenly turned up again, close at hand, at a time when it was too late to use them. The final proofs of *The Post Card* had already been sent back to the printer. "Telepathy" itself is an elaborately redacted version of the original documents, elided by "triage [*tri*], fragmentation, destruction" (*ibid.*). It would be interesting to see what got left out. In a similar way, Freud destroyed or "accidentally misplaced" the letters to him from Fliess. To compare a lesser thing with a greater one, this present fake lecture, already far too long to be read in one *séance*, ought to have been part of my *For Derrida*, but has been written too late for inclusion there.

The *Fort/Da* sequence of his losing and then refinding "Telepathy," says Derrida, "remains inexplicable for me even to this day. These cards and letters had become inaccessible to me, materially speaking at least, by a semblance of accident [*une apparence d'accident*], at some precise moment" (Te, 423; Tf, 237). He then "rediscovered them very close at hand [*tout près de moi*], but too late" (*ibid.*). Derrida claims he incinerated the originals in "tongues of fire," at least the personal parts that would, perhaps, have given away the addressees' identities. It would be interesting to know what remains, remainders, *restants*, remain at Derrida's house outside Paris, or in the archives at the University of

California at Irvine, or in the other archives at IMEC (Institut pour la Mémoire de l'Edition Contemporaine) in Normandy. Did he really burn everything?

Can we believe the story Derrida tells? Derrida's language is defensive, tortuous, and embarrassed, to say the least. Why does he say that the cards and letters were only "materially speaking at least" (*matériellement au moins*) inaccessible to him. In what way were they therefore still accessible, if only "materially" inaccessible? What is the difference between a real accident and "a semblance of accident"? Derrida knows beforehand what a reader is likely to say. He tries to ward off our suspicion that the manuscript was, for some reason, lost "accidentally on purpose." "There will perhaps [perhaps? (JHM)] be talk of omission through 'resistance' and other such things. [What 'other such things'? (JHM)] Certainly, but resistance to what? To whom? [*à quoi? à qui?*] Dictated by whom, to whom, how, according to what routes [*voies*]?" (Te, 423; Tf, 237).

Derrida here throws up a not altogether effective smokescreen. He doth protest too much. Derrida uses in his apologetic explanation terminology essential in "Telepathy" itself. That terminology also pervades the parent from which it was separated, like an orphan, "Envois." An example is the play of personal pronouns, here reduced to "whom." This vocabulary makes resistance somehow an intersubjective matter. It is resistance to something coming from a distance, by tortuous routes, perhaps telepathic ones. That is the case, even though resistance may be by way of an internal doubling of the self into an "I" and a "you," each at some distance from the other. I shall return to pronouns in "Telepathy." Derrida uses here the figure, essential to this work and to the parent work, of the paths by which a message is delivered (or not delivered) by means of a telecommunications network. That network allows many possibilities of *destinerrance*, or of *adestination*, or of becoming no more than a "dead letter." One of the letters Derrida sends to his anonymous beloved in the "Envois" to *Poste Restante* (General Delivery) apparently becomes such a dead letter, lost in the repository for such letters in Bordeaux. This happens according to a "paradox" about telepathy that Derrida formulates early in "Telepathy." He makes use, you will note, of another of his performative invocations of an indeterminable "you" or *tu*:

> For here is my latest paradox, which you [*tu*] alone will understand clearly: it is because there would be telepathy [*parce qu'il y aurait de la télépathie*] that a postcard can always not arrive at its destination. The ultimate naïveté would be to allow oneself to think that Telepathy guarantees a destination that "posts and telecommunications" fail to assure. On the contrary, everything I said about the postcarded [*cartepostalée*] structure of the mark (interference [*brouillage*], parasiting, divisibility, iterability, and so on [the last phrase is in English in the original (JHM)] is found in the network [sur le réseau]. This goes for any tele-system [*système en télé*]—whatever its content, form, or medium [*support*].
> (Te, 239; Tf, 249–50)

Though this passage appears within "Telepathy," it offers a proleptic explanation of why the letters and cards were lost "by a semblance of accident." It was in the cards, *cartepostalé*, that they would be lost, whatever the "you" to whom it was addressed, whether that *tu* was another person, the reader, or some other component of "Derrida" himself. Getting lost was "Telepathy'"s destiny, or adestiny. The series of nouns beginning with "structure" is a characteristic Derridean cascade of allusive terms metonymically displacing one another in a sideways slide. These words are nodes or switching points within the immense network that makes up Derrida's work "as a whole." Two things happen in the formulation of Derrida's telepathic paradox:

1. Telepathy becomes defined as not just a direct thought transference from one person to another, but as the entry of a thought

message into a "network" that is compared, in a way that echoes Freud, to "tele-systems" generally, whether the postal system, the telephone system, or the present-day email and Internet system. That means any telepathic message can get lost along the way or get picked up by the wrong person. Though Derrida in "Telepathy" and "Envois" stresses the immense, epochal differences a shift from one tele-system to another make, here he says all these systems, "whatever [their] content, form, or medium," are alike.

2. They are alike in being subject to the law of *destinerrance* that Derrida had worked out in his putdown of Lacan's reading of Poe in "Le facteur de la vérité,"[41] and to the law of iterability that he proposed in his revision of speech act theory in *Limited Inc*, not to speak of their submission to what his earlier work says about the "mark" as a "trace" within a system of *différance*. *Destinerrance* means that because a letter can always not reach its destination, in some sense it never reaches its destination. Iterability means the disabling of the speech act theory proposed by J. L. Austin and iterated in reductive form by John R. Searle. Any speech act enunciation can function in innumerable contexts, therefore its effect in a given context can never be predicted or controlled. Derrida's idea that I can become the person an intercepted postcard asks me to be is a special version of this. The result is that "Telepathy," just because "there would be telepathy," was always in danger of becoming a "dead letter," according to the "paradox" asserted within the text. Even when the text is suddenly found and then published, it enters a quasi-telepathic communications network of *destinerrance*, of *fort/da*, where it remains in circulation to this day, and beyond, in the future, still seeking its destination, you or me.

I have been taught, however, that resistance is a solitary matter, part of the internal structure of the self: ego, id, and superego. So why does Derrida ask "resistance to whom"? No doubt, I admit, resistance often involves reluctance to speak of something to another person, to confess something, or to ask some favor.

Nevertheless, I had thought resistance was primarily an internal matter, the self resisting itself.

I can easily think of various kinds of unconscious resistance Derrida might have had to including "Telepathy" in "Envois." He may have been embarrassed, as Freud was, by the need to face directly the question of whether or not he believed in telepathy. He may have unconsciously felt that he had treated Freud too harshly and ironically in his readings of Freud's telepathy essays. He may have unconsciously felt that "Telepathy" gives away the secret of Derrida's models for "Envois," that is, Freud's letters to Fliess and other correspondents, Flaubert's letters to Louise Colet, or Kafka's letters to Milena. He may even have felt that in various ways "Telepathy" makes too explicit the theoretical and literary presuppositions of "Envois." He may have unconsciously felt that the material in "Telepathy" really belongs in another book, perhaps the one he says, in the untitled preliminary words to "Envois," that he has not written: "You might read these *envois* as the preface to a book that I have not written" (PC, 3; CP, 7). You might.

The reader might be tempted to think of "Telepathy" and "Envois" as entirely fictive. It might be a brilliant postmodern novel in letters that Derrida made up out of whole cloth. This would make these works something like Freud's "fake lectures" on telepathy, or like Browning's fictive invention of an episode in Boston in which, so Browning maliciously imagines, D. D. Home was found out and had to talk fast to get himself out of the jam he was in. (There is jam again! What is the relation between jam as condiment and jam as a name for being caught in a shrewd situation? Maybe jam is called "jam" because its ingredients are all jammed together in the same sweet mixture.) Hardly any segments of Derrida's "Telepathy" or "Envois" would fit on a postcard, even though Derrida's speaker, early in "Envois," claims to be writing on a pile of postcards and then posting them in an envelope. Maybe so.

If the dated segments in "Telepathy" are actual love letters they are exceedingly strange ones. Long fake lectures about Freud's belief in telepathy are hardly a usual means of seduction, nor are reproaches to the beloved for being "determined" to resist him, even though all this is punctuated by brief avowals of eternal love, and repeated questions that indicate his submission to his beloved: "So, what do you want me to say?" (Te, 226, 233, etc.; Tf, 233, 244, etc). In the parent text, "Envois," the speaker says repeatedly, "I accept" (e.g. PC, 26; CP, 31). Accept what? He never says. Apparently some condition the "you" has set. The fake letters are often too long for an ordinary letter such one might write to one's beloved, much less short enough to fit on a postcard, though I suppose the many pages or cards might be stuffed into a big manilla envelope.

The circumstantiality in the text of "Telepathy" as to dates and names of "real people," however, gives one pause. Why tell the reader exactly when a given segment was written if that was not the case? Maybe just to give a bunch of lies a spurious verisimilitude? Moreover, the claim that all of "Telepathy" was written between July 9 and July 15, 1979, is more than a little implausible. That is so, however, only until one remembers Derrida's fantastic inventiveness and the unbelievable speed with which he composed essays, books, and seminars: a month or less for *Specters of Marx*, for example. How could he do it?

I could testify under oath, moreover, that what the speaker says happened the three or four times I am mentioned by name in "Envois" really did happen as "historical events." Paul de Man and I did year after year meet Derrida's plane at Kennedy when he came to give his seminars at Yale, and he did disappear briefly each year to make a phone call from a nearby booth. We never asked him whom he was calling. None of our business (PC, 107–8: September 22, 1977; CP, 118; PC, 155: September 24, 1978; CP, 168). I did take Derrida sailing on Long Island Sound, one year when he was at Yale, in my eighteen and a half foot Cape Dory Typhoon, the *Frippery*. Derrida did not know, because I never told him, that the "Small Craft Warnings" were up and that we probably should not have been out sailing (PC, 166: October 6, 1978; CP, 180). Look what happened to Shelley when he went sailing on a day when there was too much wind! Derrida and I did go together, as the "Envois" report, on another occasion, to visit Joyce's tomb in the Zürich cemetery near the zoo. The animal cries from that zoo appear in *Finnegans Wake*. We did stand laughing before the tomb of Egon Zoller, "*Erfinder des Telephonographen*," with its engraved ticker tape machine and its carved Alpha and Omega. Derrida, as we stood looking at the tomb, connected it to his then current writing about telecommunication networks, that is, the "Envois." He asked me to take a photograph later and send it to him, which I did have a friend do. It may be among his remains. It is the case, as the "Envois" say, that we searched for Peter Szondi's tomb, but failed to find it, though I found it easily enough on another visit to that cemetery. I do not remember, however, having a car in Zürich, so I do not see how I could have driven him up to the cemetery, as he asserts. Maybe I have forgotten, or maybe I rented a car for the occasion (PCe, 148: 20 June 1978; CP, 160-1). Derrida's extended story about the fall from his son's skateboard and his injured ankle, with its cross-reference to the passage about walking with a limp, at the end of Freud's *Beyond the Pleasure Principle*,[42] is historically accurate. At least that was the explanation he gave for the limp he had when he came to give his seminars at Yale that year (PC, 138, 139, 147: 20 April, 4 May, 15 June 1978; CP, 150, 152, 160). It is the case that Derrida briefly took up running at Yale in the campus cemetery, the Grove Street Cemetery, where so many Yale worthies sleep peacefully under the sod. Derrida jogged under the tutelage of James Hulbert, the translator of "Living On/Border Lines," in *Deconstruction and Criticism*. Hulbert would stop Derrida every hundred yards or so to take his pulse. He did this in order to be sure Derrida was not over-exerting himself, and was not about to join the crowd of those nearby buried ones. It must have been a funny scene (PC, 157: 26 September 1978; CP, 170). Cemeteries seem to reappear in these episodes!

Perhaps, therefore, "Telepathy" and "Envois" are throughout real letters and postcards that actually did get sent and that are based through and through on "reality." Perhaps, however, what I have just said is lies, perjury, proffered in order to provide cover for my dead friend's prevarications. Testimony, as Derrida said in his seminars on "Witnessing," is performative, not constative. You have to believe it or not believe it, without any ultimate possibility of verification. The bottom line, I hold, is that the question of these post cards' and letters' veracity is deliberately undecidable, as are their authors and their addressees. You do not know, and you will never know. You can know nothing about it. You must believe— or not believe. Even if some of the events are "real," they may have been cleverly incorporated into an overall fiction. If the medium is the maker, Derrida has made something exceedingly strange by incorporating words for these historical events into the conventions of the "Envois." The word, after all, means "sendings," as in "sending someone up."

If Derrida is right in an important passage in "Telepathy," however, the questions I have asked about the historical veracity of "Telepathy" and "Envois" are naïve, old-fashioned, just plain wrong. Psychoanalysis, says Derrida, or perhaps it is Freud speaking through Derrida (I shall return to this puzzle), has eliminated the distinction between internal events and external events. Freud has thereby put out of court, transcended, sublated, *aufgehobene*, the distinction between inside and outside, between an event that occurs in a dream and an event that occurs in the so-called external world. Therefore, it would follow, to ask whether Derrida and I did actually go, in the flesh, to visit Joyce's tomb, with its charming little statue of Joyce, is barking up the wrong tree, putting your finger in the wrong jam. That event exists, now that Derrida is dead, as dead as Joyce and Szondi, only in my, perhaps fallacious and retouched, memory of it, as well as in Derrida's words on the page and in my words in this fake lecture, which you can believe or not, as you choose. When I am gone, *fort* not *da*, or perhaps *da* but ga-ga, the event will have only a spectral, textual existence. Here is part of what Derrida says, or what he imagines Freud or Derrida/Freud as saying:

Previously, I am going back still, I had recalled that the psychoanalytic interpretation of dreams lifts up [relève], suppresses and preserves *(aufhebt)* the difference between the dream and the event *(Ereignis)*, giving the same content to both. In other words, if there should one day be someone of either sex to follow me, to follow what I still hold back in the inhibition of the too soon [i.e. his belief in telepathy (JHM)], it will be to think: from the new thought of this *Aufhebung* and this new concept of the *Ereignis*, from their shared possibility, one sees the disappearance of all the objections in principle to telepathy. The system of objections rested on a thousand naïvetés with regard to the subject, the ego [*du moi*], consciousness, perception, and so on, but above all on a determination of the "reality" of the event, of the event as essentially "real"; now that belongs to a history of grandad's philosophy [*la philosophie du grand-papa*] [The reference is either to Freud himself or to Freud's correspondent, his double by transference, who dreamed that his wife had twins, when it was his daughter who actually did so. He did not understand that the two events, the dreamed one and the real one, were "the same" from the perspective of the unconscious, but Freud straightens his readers out about that. Or the *grand-papa* could be, by condensation, both. (JHM)], and by appearing to reduce telepathy to the name of a psychoanalytic neopositivism, I open up its field [so as to incorporate telepathy within psychoanalysis (JHM)].
(Te, 253; Tf, 263)

What is at stake in this making equivalent of dream and reality? The bottom line is its support for Derrida's claim that dreams and other foretellings are performative, not constative. They bring about the event they predict. They *are*, already, in a future anterior, the event they prophesy, as my initial citation affirms.

So much for the assumption that we must decide whether the letters and postcards in "Telepathy" and "Envois" are real or fake! Derrida, however, identifies a particular and all-important undecidability in the pseudo-preface to "Envois." Let me approach what he says by way of the first sentence of "Telepathy." Everything I am writing in this fake lecture on mediums and media can be seen as a (failed) commentary on this one sentence: "So, what do you want me to say?" (Te, 226). In French: *Alors, que veux-tu que je te dise?* (Pf, 237). You will note that the "you" in the French original is in the second person singular, the tender and intimate "thou." "So what dost thou want me to say?" This use of the second person singular is the norm throughout "Telepathy." It is difficult to keep this in mind when reading the English translation. Derrida, so far as I remember, never *tutoyer*-ed me during the many years of our friendship, even in letters written in French. This was an indication, I suppose, that the distinction was not trivial for him. However, he consistently addresses whomever it is he is writing to in "Telepathy" as "thee." In a marvelous, as yet unpublished, seminar on the phrase *je t'aime* ("I love you"), Derrida stressed the importance and the difficulty of translating into English the second person singular in the French declaration of love.

The problem begins when "you" try to decide just who that "thee" is, and just who the "I" or *je* is, in the initial sentence of "Telepathy," and thereafter. The short answer is that you cannot decide. Derrida spectacularly manipulates, throughout "Telepathy," the indeterminacy of the referents of personal pronouns. Linguists do not call them "shifters" for nothing. I say "I," but you also say "I," and one cannot tell the difference by way of the word itself. Nor can one tell from the word itself the referent of a "you" or *toi*. Personal pronouns are in this the exact opposite of proper names, which presumably name one unique and unrepeatable person. "I, J. Hillis, am writing this," and "J. Hillis Miller is writing this," say two entirely different things. The first sentence might have been said by anybody anywhere at any time. The second could only be said about, well, about whom? About "me," whoever that is.

Many examples of the vertiginous effect that can be achieved by using unanchored pronouns punctuate, strategically, "Telepathy." An example is the following: "what did she want to give me or take away from me in this way, to turn away from him or in view of him [*que voulait-elle me donner ou me retirer ainsi, détourner de lui ou en vue de lui*], I don't know and I don't much care [*je m'en fous un peu*], what followed confirmed me in this feeling" (Te, 231; Tf, 242). Who in the world is saying this? Who is "she," and who is "him"? What in the world is "Derrida," who presumably wrote these words, talking about? I know nothing about it. Hegel, in a passage commented on by Paul de Man, already knew about these problems. Hegel played on the indeterminable relation in German between *mein* ("mine") and *meinen* (mean). When I say "mine" I can mean anybody's.[43]

Lewis Carroll also knew about the dizziness of multiplied personal pronouns. He exploited this vertigo brilliantly in the trial scene at the end of *Alice in Wonderland*. The context shows that Carroll knew, as did Derrida later in the passage just cited, that the shiftiness of personal pronouns has a deep relation to questions of gifts, ownership, guilt, responsibility, punishment, secrets, and truth. The poem in *Alice* has no addressee and no signature, like the letters and postcards in "Telepathy." The king says this is all the more proof that the Knave of Hearts, who has been accused of stealing the tarts (jam again), must be guilty: "You *must* have meant some mischief, or else you'd have signed your name, like an honest man."[44]

I cannot resist citing the whole wonderful poem. In my present context, it sounds like a wild rhyming parody of Derrida's "Envois":

They told me you had been to her,
    And mentioned me to him:
She gave me a good character,
    But said I could not swim.

54

He sent them word I had not gone
   (We know it to be true):
If she should push the matter on.
   What would become of you?

I gave her one, they gave him two,
   You gave us three or more;
They all returned from him to you,
   Though they were mine before.

If I or she should chance to be
   Involved in this affair,
He trusts to you to set them free,
   Exactly as we were.

My notion was that you had been
   (Before she had this fit)
An obstacle that came between
   Him, and ourselves, and it.

Don't let him know she liked them best,
   For this must ever be
A secret kept from all the rest,
   Between yourself and me.[45]

"Gave him one" what? What is the referent of "it"? "Gone" where? Just what is the "matter" the poem mentions? The uncertainty about the referents of the pronouns contaminates the reference of all the other shifter words like "it" and "one," until the poem achieves that hyperbolic degree of sensible "nonsense" in which Carroll excelled. What is this poem "about"? I answer that it is a brilliant poem about the pronoun as shifter with indeterminable referent.

In case you may think I have invented the idea that Derrida exploits the indeterminacy of pronouns in "Telepathy," or that my comparison with Carroll's poem is extravagant and implausible, or, perhaps, that I am too bad a reader to figure out just who are the "I's," the "you's," the "he's," and the "she's" in "Telepathy," I turn now to what Derrida (I guess it must be Derrida) says in the untitled "Preface" to "Envois." "Who is writing?" asks whoever is writing this. "To whom? And to send, to destine, to dispatch what? To what address? Without any desire to surprise, and thereby grab attention by means of obscurity, I owe it to whatever remains of my honesty to say finally that I do not know. Above all I would not have the slightest interest in this correspondence and this cross-section [*découpage*], I mean in their publication, if some certainty on this matter had satisfied me" (PC, 5; CP, 9).

"Come on, Jacques," the reader is likely to exclaim in exasperation. "Surely you as sovereign author know to whom all those pronouns refer, even though you may for strategic reasons or for reasons of *pudeur*, reticence, be keeping the referents secret." "No," says "Derrida," in effect, "I know it will seem scandalous, but I really do not know who is speaking and to whom. I swear I am ignorant. I assure you also that this lack of certainty is the only thing that interests me about 'this correspondence.'" He goes on then to draw the dizzying consequences of this uncertainty: "That the signers and addressees [*destinataires*] are not always visibly and necessarily identical from one *envoi* to the other, that the signers are not inevitably to be confused with the senders, nor the addressees with the receivers, that is with the readers (*you* for example [toi *par example*]), etc.—you [*vous*] will have the experience of all this, and sometimes will feel it quite vividly, although confusedly. This is a disagreeable feeling that I beg every reader, male and female, to forgive me. To tell the truth, it is not only disagreeable, it places you [*vous*] in relation, without discretion, to tragedy [*avec de la tragédie*]" (PC, 5; CP, 9).

Apparently Derrida (whoever "he" is) is asserting that it would be a tragic situation if we could not, as in this case, know who is writing and to whom, or whether a post card I write will ever reach its destination, without being purloined along the way. Why call it tragic, rather than, say, "catastrophic"? Why say *avec de la tragédie*, which I suppose might be translated as "with something of tragedy," rather than "in relation . . . to tragedy," as the translation has it. Derrida does not really explain that odd locution. Perhaps it is because this *destinerrance* puts us all, both male and female readers of the "Envois," in the situation of a tragic hero or heroine who, for example, unintentionally kills his father and sleeps with his mother, like Oedipus, or who kills her husband, like Clytemnestra, or helps kill her mother, like Electra. The question of the Oedipus complex comes up more than once in "Telepathy" and in the "Envois."

An even more apropos "tragic" situation, however, might be father–daughter incest, or the desire for it. Though Aristotle, in the *Poetics*, does not include this among his examples of tragic events, nevertheless he emphasizes that pity and terror aroused by tragedy is a family matter: "but when the tragic acts come within the limits of close blood relationship, as when brother kills or intends to kill brother or do something else of that kind to him, or son to father or mother to son or son to mother—these are the situations one should look for."[46] I suppose Agamemnon's sacrifice of Iphigenia, the father's sacrificial killing of his daughter, might be seen as a perverse form of father–daughter incest. Father–daughter love is an essential theme in a number of Freud's examples from his correspondents in his telepathy essays, for example in the story of the man who dreamed that his wife had given birth to twins, when it was his daughter who had done so, or in the story of the childless married woman who removed her ring and went to a soothsayer. The fortune-telling medium made predictions about the children she would have when she was thirty-two. These predictions did not fit her, since she was sterile, but they fit her mother perfectly, suggesting that she wanted to have children by her father. Father–daughter incest is of course the central motif in Freud's *Studies in Hysteria*.

Derrida, in the love-letters from some one of the Derridas to some unknown beloved *tu*, in "Telepathy" says, as a somewhat strange part of his seduction: "You remember, one day I told you: you are my daughter and I have no daughter" (*Tu te rappelles, un jour je t'ai dit: tu es ma fille et je n'ai pas de fille.*) (Te, 253; Tf, 263). That last sentence sounds to me like some one of the "Derridas" speaking. Freud had daughters. Derrida had only sons. The sentence appears to be "Derrida" interpolating something in his own voice in the midst of a passage in which he lets Freud speak through him about the father whose daughter had twins: "The bonds between a daughter and her father are 'customary [*habituels*] and natural,' one should not feel ashamed of them. In everyday life, it expresses itself in a tender interest, the dream alone pushes this love to its final conclusions [*conséquences*], etc." (Te, 253; Tf, 262–3). As for a father's desire to sleep with his daughter, or of a daughter's desire to sleep with her father, I can say what Freud said about telepathy: "I have no opinion on the matter and know nothing about it" (SE 18: 220).

At the end of what I am calling the "Preface" to "Envois," Derrida assures the reader, "I assume without detour the responsibility for these *envois*, for what remains [*reste*], or no longer remains, of them, and that in order to make peace within you [as opposed to tragic inner turmoil (JHM)] I am signing them here in my proper name, Jacques Derrida" (PC, 6; CP, 10). That seems clear enough and wonderfully reassuring. Derrida admits that he did really write "Telepathy" and the "Envois," so it is all right to say, "Derrida says" this or that in these works. Unfortunately, a footnote to the fake, printed signature takes away with one hand what he has just given with the other: "I regret that you [*tu*] do not very much trust my signature, on the pretext that we might be several. This is true, but I am not saying so in order to make myself more important by means of some supplementary authority. And even less in order to disquiet. I know what this costs. You are right, doubtless we are several, and I am not as alone as I sometimes say I am when the complaint escapes from me, or when I still put everything [*je m'évertue encore*] into seducing you" (PC, 6; CP, 10). This footnote puts the reader right back where he or she has been all along. Who is talking, which one of the Derridas, and to whom is he or she talking when he or she says, "when I still put everything into seducing you"? Can it be I, the reader, whom "Derrida" is trying to seduce?

The copy of *La carte postale*, that is, the French original, that Derrida gave me in 1979 has what appears to be a genuine signature, "Jacques," written in pen below the fake signature in typeface that appears in the printed books, both French and English. After the title of the first section, Derrida has added, with the same pen and in the same hand, my name and my wife's, so that it reads *Envois à Hillis, à Dorothy*. I must say that does not entirely reassure me. In fact, it makes me distinctly uneasy. I am not sure either of us wants to be the designated addressee of a set of possibly fake letters and postcards, especially this particular set. I would rather leave that undecidable. Making us uneasy was no doubt what Derrida intended. It was his affectionate little joke.

Just what is at stake in all this foolery with names and pronouns, with indeterminate senders and receivers, with multiple personalities (*nous sommes sans doute plusieurs*)? The text suggests three not entirely compatible answers:

(1) Derrida wants "Telepathy" not just to talk about the performative power of letters, which he says the theorists of speech acts have tended to ignore. He wants the letters themselves to act performatively, to create both the sender and the addressee, in an extravagant act of seduction. This happens according to the theory he expresses so eloquently in the first pages of "Telepathy." I make another long citation here from a much longer development because what Derrida says is so central to the rhetorical strategies of "Telepathy" and of the "Envois," and therefore so central also to my reading of these strategies:

> I am not putting forward the hypothesis of a letter that
> would be the external occasion, in some sense, of an
> encounter between two identifiable subjects—and who
> would already be determined. No, but of a letter that

after the event seems to have been launched toward some unknown [*inconnu(e)*] addressee at the moment of its writing, an addressee unknown to himself or herself, if one can say that, and who is determined, as you [*tu*] very well know how to be, on receipt of the letter; this is then quite another thing than the transfer of a message. . . . So you say: it is I [*moi*], uniquely I, who am able to receive this letter, not that it has been reserved for me, on the contrary, but I receive as a present [*un présent*] the chance to which this card delivers itself. It falls to me. [*Elle m'échoit.*] And I choose that it should choose me by chance [*au hasard*], I wish to cross its path [*trajet*], I want to be there, I can and I want—its path or its transfer. In short you say "It was me," with a gentle and terrible decision, altogether otherwise [*tout autrement*]: no comparison here with identifying with the hero of a novel. (Te, 228, 229; Tf, 240)

(2) Derrida wants to suggest that one possible way of reading these fragmentary texts, a way that cannot be cleanly distinguished from the others, is as a transaction that all takes place internally. The I–thou in question may be a relation between one or more imaginary Derridas and other imaginary Derridas, in a breakdown of the distinction between inner and outer events that Derrida asserts in passages cited and commented on above. In one place, not so far cited, he suggests explicitly that both "I" and "you" are internal to "Derrida," or perhaps not, in a dizzying alternation:

And not to reply is not to receive. If, from you [*de toi*] for example, I receive a reply to this letter, it is because, consciously or not, as you wish, I'll have asked for this rather than that, and therefore from this man or that woman. As this seems at first, in the absence of the "real" addressee, to happen between myself and myself, within myself [*à part moi*, also "except for myself"], a part of myself that will have announced the other to itself [*qui se sera fait part de l'autre*; also "that will have made itself a part of the other"], I will clearly have to have asked myself . . . What is it that I ask myself, and who? You [*Toi*], for example, but how, my love, could you be only an example? You know it, yourself, [*Tu le sais, toi,*] tell me the truth, O you the seer, you the soothsayer [*la devine*]. What do you want me to say to you [*Que veux-tu que je te dise*], I am ready to hear everything from you, now I'm ready, tell me. (Te, 234–5, trans. modified; Tf, 245)

(3) Derrida wants to set the stage for the uncertainties of his reading of Freud. This reading takes the form of an intermittent dramatic monologue in which "Derrida" becomes the medium through whom Freud speaks. This transition takes place by way of pronoun shifters. At first you think it is one of the primary "I's" speaking. Then suddenly you realize it must be Freud speaking. Or rather, it must be one of the Derridas speaking as a medium imagining what must have gone on in Freud's mind as he wrote those fake lectures on telepathy. These lectures were never intended to be delivered orally, as I have said, but they sound like real lectures, with apostrophes to "Ladies and Gentlemen," and references to "today" as the time of delivery and so on: ""Ladies and Gentlemen,—Today we will proceed along a narrow path, but one which may lead us to a wide prospect" (SE 22: 31). The effect of the unsignalled transitions from "Derrida," speaking as "I" to his beloved as "thou" (really you or I as reader being invoked into existence by the words, the reader he is trying to seduce), to passages that you gradually realize must be "Derrida" pretending to be Freud, is extremely peculiar and unsettling. The speaker uses a form of language that Derrida calls "a mimetic or apocryphal style as Plato would say" (Te, 253; Tf, 262). The reference is to Socrates' stern condemnation, in *The Republic*, of "double diegesis" as immoral. In the *Odyssey*, says Plato/Socrates, Homer speaks not as himself but as Ulysses, just as Plato himself uses "double diegesis"

by speaking as Socrates condemning double diegesis. This is immoral because it breaks down the single, unique, unitary, ethically responsible self. It is "apocryphal" because the speeches are invented. They never really happened, just as Mr. Sludge never made the speech Browning ascribes to him.

A given entry in "Telepathy" will start out sounding like it must be one of the "primary" speakers addressing his beloved, in English, as "My sweet darling girl" (Te, 239; Tf, 250). But this addressee is really you or me, someone, that is, conjured up by way of "Derrida"'s invocation: "put yourself [*mets-toi*] in the place of another feminine reader, it doesn't matter who, who may even be a man, a feminine reader of the masculine gender" (Te, 229; Tf, 241). Then "you" suddenly realize, because the name "Martha" is mentioned, that it must be Freud as a young man writing to his wife-to-be, Martha: "to organize with Eli our meeting on Saturday and to smuggle this audacious missive as contraband. But it seems to me impossible to defer sending [*l'envoi de*] my letter" (Te, 239; Tf, 250). These switches are like Browning's Sludge speaking first in the fake voice of the spirit whose medium he claims to be and then lapsing back into his own voice, though his clients at the *séance* do not seem to notice or mind if they do notice. Or it is like tuning a radio receiver so first one voice comes through from what used to be called the "ether" (also of course the name of an anaesthetic) and then, as that fades, another. Radio uses the ether to hypnotize you and put you to sleep.

A prime example of this strange telepathic effect is the whole segment that purports to have been written on July 13, 1979. This segment begins with a paragraph that refers to "Safah, the name of the 'lip' [*lèvre*] and of my mother, as I told you in October" (Te, 244; Tf, 254). Derrida's mother's name was Georgette Safar. Why the "r" becomes an "h" in Derrida's notation, I do not know. *Safah* means, in Hebrew, "lip," in the sense of "tongue" or "language." The word is used in the passage about the Tower of Babel in Genesis that Derrida cites several times in the "Envois." Safah/Safar would be an example of what Derrida finds essential in the Biblical account of the Tower of Babel. This is its demonstration of the impossibility to reconcile a proper name and its conceptual homonym. YHWH exemplified the confusion of tongues (or "lips"), even within each separate language, by giving the city with its tower the proper name "Babel," or "Bavel," in another shifting of consonants. "Bavel" means "confusion," as the Bible says, so it is both a proper name and a common word, in inextricable confusion, just as Safar(h) is Derrida's mother's and maternal grandfather's surname, and at the same time a common noun meaning "lip" in the sense of language. [47]

In any case, this must be "Derrida" speaking. Or perhaps, in a Babelian confusion of tongues or lips, it is Freud and Derrida superimposed and speaking at once about how they "were making one read under hypnosis" (Te, 244; Tf, 254). Both Derrida and Freud do that. The next paragraph, however, is definitely Freud speaking telepathically through Derrida: "and so in my first period, that of indecision. In the fake lecture entitled 'Dreams and Telepathy,' my rhetoric is priceless, really incredible. Incredible, that's the word, for I play on credibility or rather acredibility, as I did a short while ago in *Beyond*" (Te, 244; Tf, 254). The reference is to the way Freud, toward the end of *Beyond the Pleasure Principle*, says: "It may be asked whether and how far I am myself convinced of the truth of the hypotheses that have been set out in these pages. My answer would be that I am not convinced myself and that I do not seek to persuade other people to believe in them. Or, more precisely, that I do not know how far I believe in them" (SE 18: 59). Slippery fellow, that Sigmund Freud, as slippery as Jacques Derrida! The apparatus remains tuned in on Freud's voice for the next page and a half. Then that voice suddenly fades and we are hearing "Derrida" express his thoughts about lying: "Do you believe that one can talk [*qu'on puisse parler*] about lying in philosophy, or in literature, or better, in the sciences? Imagine the scene: Hegel is *lying* when he says in the greater *Logic* . . . or Joyce, in some passage from *F. W.*, or Cantor? but yes [*mais si*], but yes, and the more one can play [*on peut jouer*] at that, the more it interests me" (Te, 246; Tf, 256). *On,*

French for "one," meaning someone, anyone, is much more idiomatic in French than in English. It is the locus, as in this passage, of a shifting indeterminacy like that of personal pronouns. "One can talk . . . "; "one can play." Who is this "one"? No way to know. The passage then reverts to Freud's imagined confessional and apologetic voice.

In this strange unacknowledged dramatic monologue, "Derrida" imagines Freud justifying his rhetorical procedures, the way he talks autobiographically, for example, about his daughter Sophie's death, while pretending to be analyzing the letters about telepathic anticipations of death or birth he has received from correspondents not "personally known" to him. This apocryphal "Freud" compares his procedures in his fake lectures to those of a spiritualist medium: "From the moment I [i.e. Freud (JHM)] started talking about hypnosis and telepathy (at the same time), a long time ago now, I always drew attention to the procedures of diverting attention [*détournement d'attention*], just like 'mediums' do" (Te, 249-50; Tf, 259). In this extended apostrophic apology "Derrida" imagines Freud to be speaking from some undetermined place and to some undetermined auditor, you or me, dear reader.

"Derrida" stresses repeatedly four elements of Freud's essays on telepathy:

(1) One is the way Freud's writing was a transfer from his early practice of therapeutic hypnosis. Now he puts his readers to sleep, so that they do not know what rhetorical tricks are being played on them: "I have never been able to give up hypnosis, I have merely transferred one inductive method onto another: one could say that I have become a writer and in writing, rhetoric, the staging of and composition of texts, I have reinvested all my hypnogogic powers and desires. What do you want me to say, to sleep with me, that is all that interests them, the rest is secondary" (Te, 248; Tf, 258).

(2) Derrida has Freud repeatedly stress the way psychoanalysis can subdue, encompass, retain power over, telepathy by breaking down the distinction, in a set of presumptions already noted above, between an unconscious or mental event and a real one, or between

dream and waking: "To leave to decide, that's the great lever, I try to place the fictive listener, in short, the reader in the situation of the dreamer where it's up to him to decide—if he's sleeping" (Te, 249; Tf, 259). The "fictive reader": that is you or me. Do I wake or sleep? I know nothing about it. The breakdown of these distinctions occurs in the claim that dreams and telepathic messages are performative, not constative, as other passages, cited above, have also asserted:

> An annunciation can be accomplished, something can happen [*peut arriver*] without for all that being realized. An event can take place that is not real. My customary distinction between internal and external reality is perhaps not sufficient here [as in the impossibility to know whether the "you" "Derrida" apostrophizes is someone, some other person, outside himself, or no more than another of the "several selves" he is (JHM)]. It [the annunciation] signals toward some event that no idea of "reality" helps us think. But then, you will say [*diras-tu*], if what is announced in the annunciation clearly bears the index 'external reality,' what is one to do with it? Well, treat it as an index, it can signify, telephone, telesignal another event that arrives before the other, without the other, according to another time, another space, and so on. That is the *abc* of my psychoanalysis. (Te, 248; Tf, 258)

(3) Derrida iterates the important recognition that the "material" Freud presents in his essays on telepathy, the material evidence, is all letters, postcards, telegrams, and telephone calls: "Once again it is a letter that reassured me. In the introductory part [*protocole*] of the lecture, already, a letter and a postcard come to refute the telepathic appearance of my two dreams—that ought to have troubled the reader. Then in the two cases described the post again officiates: two correspondents who are not 'personally' known to me" (Te, 251; Tf, 261).

(4) Finally, Derrida has Freud stress throughout the importance of ghosts and premonitions of death in all these telepathic postal stories. This imagined Freud also stresses the obscure relation between that emphasis on death and the "primal scene" of the infant Freud's fall from a stool when he was reaching up to get at some forbidden food, probably jam. He bears the scar (*Narbe, Spur*) to this day, says Freud to his imagined audience. He can part his beard to show the scar. The wound, Derrida seems to be implying, was a symbolic castration: "The word *Narbe* [scar] comes twice from my pen, I know that the English had already used the word 'scar' to translate *Spur* much earlier on. This translation may have put some people on the trail [*piste*]. I like these words *Narbe*, 'scar,' *Spur*, trace, *cicatrice* in French as well. [This must be Derrida's voice momentarily drowning out Freud's. (JHM)] They say what they mean, eh, especially when it is found under the bristles of some *Bart*, or beard. Nietzsche already spoke about the scar under Plato's beard. One can stroke and part the bristles so as to pretend to show, that is the whole of my lecture" (Te, 251; Tf, 260–1). What Freud pretends to show is an absence, not a presence, with an implied allusion to castration. It is a wound that has never really scarred over, but that, "don't you think [*n'est-ce pas*], opens the text, holds it open, mouth agape, the analytical material come from elsewhere, in my dossier on telepathy, remains through and through *epistolary*" (Te, 246–7; Tf, 256–7).

I conclude this section with the claim that "Telepathy" is, among other things, a superb example of Derrida strategy of critical or "deconstructive" reading. It is a deconstructive reading that uses as its primary strategy a diabolically ironic miming of Freud's essays on telepathy. This miming uses as its tool a "micrological" reading of a fake dramatic monologue, that is, the one Derrida imagines Freud speaking. This is a species of du Bosian introspection of the other. It is not introspection of the other's consciousness, which is what Charles du Bos meant. Derrida thinks that is impossible, or he sometimes thinks he thinks it is impossible, in spite of the way he also finds "non-telepathy" impossible to imagine. The consciousness of the other remains, for Derrida, on one side of this fold, following Husserl's reluctant conclusion, inaccessible, secret, wholly other, except by what Husserl called, in the fifth Cartesian Meditation, "analogical apperception," a phrase repeatedly cited by Derrida in various works.[48] That indirect, analogical access works, for Derrida, by way of the language of the other. That language *is* accessible in the quasi-telepathic way we call "reading," that is, by introspection of the language of the other.

Reading as ironic miming both claims telepathic access, and at the same time keeps at a distance, undermines by miming, by repeating or paraphrasing just what Freud said, in the first person. *Fort/Da*; Freud/Derrida; *tele*pathy/tele*pathy*. This miming convicts Freud, by way of his own words, of saying, in covert autobiographical self-revelation, more than he intended to say. He intended, rather, to lull his readers to sleep, to divert their attention, by a process Derrida calls "hypnopoetics" [*hypno-poétique*] (Te, 253; Tf, 263). This is poetics as a performative means of hypnotizing the reader, putting her or him to sleep, in a reverse or perverse form of "making."

Derrida, however, though he often claims in "Telepathy" to be asleep, or just waking from sleep, is in the end too wide awake to be fooled. He claims, for example, to recognize that Freud's account, in "Dreams and Telepathy," of the material about telepathy that reaches him by correspondence is not there for its own sake. It serves as a smoke-screen strategy that works to offer an oblique reading of the two dreams of his own with which he begins. These dreams are ostensibly presented to demonstrate that he, Freud, has never had a telepathic premonitory dream that has been fulfilled, *wie wir uns ausdrücken, eingetroffen,* "not themselves, as we put it, realized," or "come true" (Te, 248; Tf, 257). Derrida goes on to say, however, that though "*eintreffen* does mean, in the broad sense, 'to be realized,'" he would prefer to translate it as "to happen" [*arriver*], meaning that "An annunciation can be accomplished, something can happen without for all that being realized. An event can take place that is not real" (Te, 248; Tf, 258). "The material that follows," says Derrida, speaking as Freud and about Freud's "Dreams and Telepathy," "and that reaches me by *correspondence*, it's sufficient to be vaguely alert or sophisticated [*déniaisé*] to understand it: it is there only in order *to read* my two dreams of death, or, if you prefer, so-that-not, in order not to read them, in order, on the one hand, to divert attention from them, while on the other paying attention to them alone" (Te, 249; Tf, 259).

No moderately wide-awake reader of Derrida's "Telepathy," however, if he or she has managed not to have had his or her attention diverted or not to have been hypnotized, lulled into sleep by the twists and turns of Derrida's annunciations, can doubt that, by a form of telepathic transference or identification the essay itself talks about, Derrida is covertly writing his own "auto-bio-thanatography" (Te, 260, trans. modified; Tf, 270). This is analogous to what Freud did in the telepathy essays. Derrida is obliquely, by transference, *Übertragung*, telling his life story as it was defined throughout by his obsession with death, especially suicide by drowning. This obsession with death extends from that adolescent poem cited in *Glas* and used by Valerio Adami in a lithograph/poster already mentioned: *glu de l'étang lait de ma mort noyée*, through the "Envois," and down to Derrida's last seminars and that last interview in *Le monde*, written in the shadow of his impending death.[49] Derrida, in "Telepathy," is revealing himself and analyzing himself at the same time.

This transference and counter-transference is made particularly evident by the way the imaginary dramatic monologue in which Freud speaks is so often mingled with, parasitized by, interrupted by, words that sound as if they must be "Derrida himself" speaking. I use "parasite" in the technological sense to name the way that a radio message is interfered with by an alien signal, ghostly music or another voice coming on a different frequency. An example of such parasiting occurs when "Derrida" says, suddenly, in an interpolated broken phrase, bounded on each side by the fifty-two spaces of ellipsis and outside them two pieces of language, one before, one after, that are clearly the imaginary Freud speaking: "It is so long since I wrote that to you [*que je t'ai écrit ça*], I no longer know" (Te, 249; Tf, 259).

In "Telepathy" Derrida sees and exposes Freud's sleight of hand, just as Browning claimed to do with Home in "Mr. Sludge, 'The Medium,'" and just as, in my much more modest way, I am claiming to do with Browning, Freud, and Derrida, mediumistically miming them ironically. I am therefore gently, ever so gently, undercutting them in my turn. Can it be, however, that I am unwittingly, in my sleep as it were, by a species of unconscious transfer, writing my own auto-bio-thanatography, as some people (not me, of course) think all so-called critical reading is? I know nothing about it.

Two more features of Derrida's "deconstructive" reading of Freud in "Telepathy" must be identified, in conclusion of this last section before my coda. One is Derrida's choice of minor and obscure works by Freud. Derrida, as is customary for him, beginning with his use of Rousseau's "Essay on the Origin of Languages," in *Of Grammatology*, continuing through his use of Hegel's family letters in *Glas*, and so on, reads four essays by Freud on telepathy, usually considered to be minor and slightly embarrassing parts of his work. He uses these to suggest a global reading of Freudian psychoanalysis that would radically transform our understanding of Freud. The abbreviated one-volume version of Ernest Jones's three-volume *Life and Works of Freud*,[50] omits entirely the chapter on Freud and occultism, presumably because it was not considered important. Neither "occultism," nor "telepathy" appears in the index.

Derrida's procedure might be called the "loose thread" theory of reading. Find an obscure thread that does not even seem to be a part of the figured tapestry, pull on it a little, and the whole figure in the carpet will unravel.

To identify the other reading strategy will unfold what I mean by "ironic miming." Much reading of philosophical or critical works, even of poems or novels, operates by moving toward abstract generalizations that the reader claims are the "kernel" of

what the text says and that can be formulated as rational proposi-
tions, e.g. "the unconscious is the discourse of the other." Derrida's
"deconstructive" reading strategy goes in the other direction,
toward "micrological" attention, such as psychoanalysts and
mediums are said to excel at. Derrida looks carefully at tiny, apparently insignificant, details of language or gesture. These are
usually, in one way or another, tropes. The background for this
attention to detail might be defined as the assumption of gross or
large-scale forms of transference, *Übertragung*, metaphor, carrying
over. Freud's "Dreams and Telepathy" appears to be about some
telepathic dreams Freud's "correspondents" sent him by letter, but
Freud's discussion of these are actually covert readings, by transference, of Freud's own discounted dreams presented and dismissed
at the beginning of the essay. The essay, Derrida asserts, is really
about Freud. It is a "hypnopoetic," "auto-bio-thanatographic" self-
reading by way of a large-scale trope of substitution that
psychoanalysis calls "transference."

Often the small-scale focus, which this large-scale assumption of
transference enables, goes by way of making figures of speech
salient that a more abstract reading might assume are just there as
superficial embroidery. They can easily be taken as no more than a
"vivid" way of expressing conceptual doctrines. Derrida thinks
otherwise. He thinks the real story is told by those figures as they
link one part of the text to another, and as they are embedded in
the idiomatic specificity of a unique speaking or writing situation.
It is a big mistake, Derrida believes, ever to rise above these partic-
ularities to some abstract conceptual level. One example is all those
odd appearances in Freud, and then in Derrida's commentary, of
references to jam, jam at the center of the earth, *der Erdkern aus
Marmelade besteht* (cited Te, 259; Tf, 269); jam as the cause of
Freud's scar; jam prepared in his correspondent's dream by his wife
who has just given birth to twins and then feeds the jam to those
newborns.

Another, and final, example, is a somewhat hallucinatory and
devastatingly comic passage that "reads" one of Freud's letters to

Jones. The letter both says Jones should tell everyone in England, for diplomatic reasons, that Freud's belief in telepathy is a private and contingent aspect of his life, like his being Jewish or like his fondness for cigars, and at the same time says belief in telepathy is now a big deal for him, something about which he feels obliged to come out of the closet:

> Even if one takes into account what he says about "diplomacy" and the diplomatic advice that he again gives to Jones, this letter is contradictory from start to finish. Enough to make one lose one's head, I was saying to you the other day, and he himself once declared that this subject "perplexed him to the point of making him lose his head." It is indeed a question of continuing to walk with one's head under one's arm ("Only the first step is costly" ["*il n'y a que le premier pas qui coûte*"], etc.) or, what amounts to the same thing, of admitting a foreign body into one's head, into the ego of psychoanalysis. Me, psychoanalysis, I have a foreign body in my head (you remember [*tu te rappelles*]. (Te, 258; Tf, 267–8)

As you can see, Derrida does no more here than string together, highlight, and, so to speak, reify, those apparently extraneous metaphors in Freud's discourse, figures of heads and foreign bodies. These would, in a more solemn reading, melt away in the construction of conceptual formulations. The effect of Derrida's procedure is devastating to any project of conceptual unification about what "Freud said," and also very funny. Deconstructive reading is comic miming. It takes nothing away from the text and adds nothing to it, except irony, as in the insolence of a student who repeats back to the teacher exactly what that teacher has said, and as in the ironic insolence of my mediumistic citations of Derrida's "Telepathy." These citations and my accompanying commentary implicitly claim to understand the essay better than Derrida did—an extremely risky claim.

Brief Coda or Post-Modern Tail

ERRIDA was one of the last to exploit the telepathic powers of the pen, the typewriter, and the telephone, just before those techniques of communication at a distance gave way, even to some degree in Derrida himself, to new techno-tele-patho-communications media: the computer, the Internet, faxes, email, and so on. (Derrida never used email, as I have said.) I am using these media at this moment to write these words, or they are using me to get themselves embodied on the computer screen and then saved to a file that may be sent anywhere in the world as an email attachment. They can then be downloaded to create the appropriate not-yet-existant addressee in whoever opens the attachment and reads.

Elisabeth Murdoch, daughter of Rupert Murdoch, is today married to Matthew Freud, great-grandson of Sigmund. Matthew Freud is a "prominent public relations executive." They have two children, more Freuds! Think of that! And think of Matthew Freud married to Rupert Murdoch's daughter! She is a power in American and world-wide television. Matthew Freud is a distant cousin of that Ernest W. Freud Derrida mentions in "Telepathy" (Te, 238; Tf, 249). He says he must still be alive in 1979 in London, as indeed he was, practicing psychoanalysis under his mother's maiden name, Freud, not his father's name, Halberstadt. "Poor sons-in-law [*pauvres gendres*]" exclaims Derrida (T, 238; Tf, 249). The French term "*gendre*" suggests that sons-in-law are good only for engendering grandchildren. They function merely as a way of avoiding the secretly desired father–daughter incest. Sons-in-law fare badly in all the epistolary stories Freud tells. The fathers and grandfathers, who do the story-telling, are jealous of them and would really rather themselves be married to their daughters and grand-daughters. The Ernst Freud whom Derrida mentions was Sigmund's grandson, son of his beloved daughter Sophie, and the hero of the famous "*fort/da*" story in *Beyond the Pleasure Principle*.[51] These connections boggle the mind. *Fort! Da!*[52]

It is wonderful to think of Freud's great-grandson at work carrying on his great-grandfather's work in the most modern way. He is not curing people of their hang-ups. He is persuading them to get hung up so they will buy certain products or believe certain things about companies or about politicians. Whatever may be the case in Britain, which is no doubt different from the United States, public relations firms in the United States work to persuade us and our elected representatives, for example, that the trade agreement with Columbia is a good thing, that John McCain is a reasonable, moderate man, who will protect the United States from all those terrorists in our midst or plotting to invade the country,[53] that solar energy will not work, or that big oil companies are doing good work for the environment, or that we *must* buy a certain drug or else die of "a dangerous clot," or of erectile disfunction, or of constipation, or of seasonal allergies, or of Alzeimer's disease, or whatever. Powerfully inventive television sales pitches by pharmaceutical companies continue unabated during the evening news even though the economy has tanked and fewer people have money enough to buy these drugs. Public relations firms do their good work by creating fictions that are taken as realities, that is, they generate ideological beliefs by means of the new media. Advertisements and news stories bring about the thing they fictionalize, just as Freud/Derrida said letters and dreams do. If we see it on television or in a magazine advertisement, it must be true. This is by no means entirely different from Freud's sleight of hand in the telepathy essays. Those essays are calculated, as Derrida says, to lull to sleep readers of lectures that were never given, just as Freud hypnotized his early patients. Perhaps he even went on to the end practicing hypnotism to get patients to dredge up their unconscious.

Elisabeth Murdoch runs "Shine," a television company, and recently (March 2008) bought "Reveille," another television production company. Both companies adopt highly successful network television shows for different countries, making them faster-paced for the United States, for example, and slower for Britain—a weird assumption! "Ugly Betty," "The Office," and "The Biggest Loser" are Reveille productions. One of Reveille's shows, "Are You Smarter Than A 5th Grader?" is broadcast, with appropriate changes, in over sixty countries. Ms Murdoch is said to have an uncanny ability to know what viewers will like, by a kind of telepathic insight, or feeling at a distance. In a legendary example of this, she urgently recommended to her father that he buy "American Idol" for Fox. It has been a gigantic success. It is even shown, in an appropriate version, in Afghanistan, or was, until a recent crackdown.[54]

"Shine," "*Schein*," "*erscheinen*," get it? As in Hegel's famous definition of art: *Das sinnliche Scheinen der Idee*. ("The sensible shining forth of the idea.") Only in Shine's and Reveille's case it is a mediumistic "idea," a simulacrum, not the Hegelian transcendent one. "Reveille" is a wonderfully antithetical name too, since the work of these television shows is to hypnotize viewers, to put them to sleep and keep them asleep. In that sleep state they will have wonderful dreams of faroff events by way of television, iPhones, iPods, all the current prestidigitizing mediumistic gadgets that make us constantly subject to touch, feeling, hearing, and seeing at a distance. We use these devices to "reach out and touch someone." The viewers of Shine's productions, carefully tailored to a given country's tastes, will, moreover, the producers hope, also go to the nearest Wall Mart or its equivalent and buy, in hypnotized sleepwalking, whatever is advertised in the "breaks" of these shows.

The purpose of dreams, Freud said, is to keep you asleep. Freud claimed to be waking you up, with the "talking cure." He brought you to remember and then to work through forgotten traumas, after, sometimes, having put you into hypnotic sleep in order to get at those buried memories. If Derrida is right, Freud's essays or fake lectures on telepathy were, nevertheless, intended to put his audiences or readers to sleep, or into a state in which they could not decide if they were awake or dreaming. This Freud did so that his hearers or readers would not notice the sleight of hand whereby his lectures talked about himself while seeming to be talking, objectively, about correspondents who were not even "personally" known to him. He also wanted to hide his belief in telepathy. Like great grandpapa, like great grandson. Matthew Freud and his wife are keeping up the good work of "producing" what they apparently just name. The medium is the maker with a vengeance in that work.

Do I believe in telepathy? Do I believe in what I as medium have allowed Browning, Freud, and Derrida to say through me? I do not know whether I believe or disbelieve. I know nothing about it.

Notes

1   Jacques Derrida, "Telepathy," trans. Nicholas Royle, in *Psyche: Inventions of the Other*, Vol. I, ed. Peggy Kamuf and Elizabeth Rottenberg (Stanford, California: Stanford University Press, 2007), 229, henceforth Te, followed by the page number; *ibid.*, "Télépathie," in *Psyché: Inventions de l'autre* (Paris: Galilée, 1987), 240, henceforth Tf, followed by the page number. I have added italics in the English that were not carried over from the French.

   Nicholas Royle's translation of "Telepathy" was initially published in the *Oxford Literary Review*, 10: 1–2 (1988), 3–42. I am grateful to the editors of this journal for allowing me to make citations from this translation.

2   See David Crystal, *Txtng: The Gr8 Db8* (Oxford: Oxford University Press, 2008) for a fascinating account of the language of texting and its social function. Can one imagine Freud sending a text message to Jung or Jones on his iPhone, complete with "emoticons" like " : ;)"— the ironic smiley? That would have been a new kind of spectral communication by way of a new medium. What difference would that have made to psychoanalysis? Barack Obama was so attached to his Blackberry that a new "securitized" one had to be provided for him after he became President of the United States. How will that change American politics?

   Barack Obama attached to his Blackberry is an example of what Jean-Luc Nancy calls the "ecotechnical." This, as my initial epigraph from Nancy's Corpus asserts, is the assumption that the whole world is an immense machine that has generated our bodies and plugged them in various ways into the "environment." Nancy's usage of the word "ecotechnical" is quite different from *Wikipedia*'s definition of ecotechnology as "an applied science that seeks to fulfill human needs while causing minimal ecological disruption, by harnessing and subtly manipulating natural forces to leverage their beneficial effects" (http://en.wikipedia.org/wiki/Ecotechnology, accessed 03/21/09). One must, to use *Wikipedia*'s own term, "disambiguate" the two usages.

3 See Nicholas Royle, *Telepathy and Literature: Essays on the Reading Mind* (Oxford: Blackwell, 1990); *ibid., The Uncanny* (Manchester: Manchester University Press, 2003), especially "The 'telepathy effect': notes toward a reconsideration of narrative fiction," 256–276; *ibid.*, "The Remains of Psychoanalysis (I): Telepathy," in *After Derrida* (Manchester: Manchester University Press, 1994); *ibid.*, "Memento Mori," in *Theorising Muriel Spark: Gender, Race, Deconstruction*, ed. M. McQuillan (London: Palgrave, 2001). Royle's brilliant essay on the "telepathy effect" is also included, in a slightly different form, in *Acts of Narrative*, eds. Carol Jacobs and Henry Sussman (Stanford, California: Stanford University Press, 2003), 93–109. This present book follows the program laid down in this provocative essay.

4 "The 'telepathy effect,'" in *The Uncanny*, 261.

5 See Pamela Thurschwell, *Literature, Technology and Magical Thinking: 1880–1920* (Cambridge: Cambridge University Press, 2001).

6 See Martin McQuillan, "Tele-Techno-Theology," in *Literaglossia*, ed. Julian Wolfreys (Edinburgh University Press, 2002), 279–97; reproduced as chapter 3 in *Deconstruction after 9/11* (New York: Routledge, 2008).

7 See Marc Redfield, "The Fictions of Telepathy," a review of Nicholas Royle's *Telepathy and Literature*, in *Surfaces*, 2: 27 (1992), 4–20.

8 See Julian Wolfreys, *Victorian Hauntings: Spectrality, Gothic, the Uncanny and Literature* (Palgrave: Houndmills, Basingstoke, Hampshire: 2002). Wolfreys' chapter on George Eliot's *The Lifted Veil* (74–93) is especially relevant to my concerns in this book.

   Christine Ferguson's *Determined Spirits: Eugenics, Heredity, and Racial Hygiene in Transatlantic Spiritualist Writing, 1848–1910* is forthcoming from Edinburgh University Press in Julian Wolfreys's Series, Edinburgh Critical Studies in Victorian Culture.

9 Freud's chief essays on telepathy are "Psychoanalysis and Telepathy," "Dreams and Telepathy," "Dreams and Occultism," and "The Occult Significance of Dreams," in *The Standard Edition of the Complete Psychological Works of Sigmund Freud*, trans. and ed. James Strachey et al. (London: Vintage Books; The Hogarth Press, 2001), 18: 173–93; 195–220; 22: 31–56; 19: 135–38. Henceforth SE, followed by the volume and page numbers.

10 http://en.wikipedia.org/wiki/The_Medium;
   http://en.wikipedia.org/wiki/The_Telephone%2C_or_L%27Amour_à_trois.

11 George Eliot users a similar figure in *The Lifted Veil* to describe the hum and buzz of her unfortunate hero's involuntary "tuning in" on others' thoughts: "the vagrant, frivolous ideas and emotions of some uninteresting acquaintance—Mrs Filmore, for example—would force themselves on my consciousness like an importunate, ill-played musical instrument, or the loud activity of an imprisioned insect" ([Harmondsworth, England: Penguin Books—Virago Press, 1985], 19). As Patience Moll, who called my attention to this passage, observes, the "loud activity of an imprisoned insect" would be interpreted by another insect of the same species as a comprehensible distress call and warning. *The Lifted Veil* and Charlotte Brontë's *Jane Eyre* are notable Victorian novels in which telepathy figures. Virginia Woolf's *Mrs. Dalloway* and Salman Rushdie's *Midnight's Children* exemplify the continued fascination with this topic in modernist and post-modernist fiction. Royle discusses these in "The 'Telepathy Effect,'" and elsewhere. He adds Charles Dickens, Thomas Hardy, Henry James, E. M. Forster, and D. H. Lawrence to the list of those English novelists who use telepathy both thematically and structurally. I would add one more, Toni Morrison's *Beloved*, to these. No doubt many others exist, in many languages. Telepathy is a big topic in literature. My focus here remains on Browning, Freud, and Derrida.

12 Ant communication is so fascinating and so contrary to what Freud imagined that I give here a summary of the latest scientific account. See http://en.wikipedia.org/wiki/Ant#Communication:

Ants communicate with each other through chemicals called pheromones. These chemical signals are more developed in ant species than in other hymenopteran groups. Like other insects, ants smell with their long and thin antennae that are fairly mobile. The antennae have a distinct elbow joint after an elongated first segment; and since they come in pairs—rather like binocular vision or stereophonic sound equipment—they provide information about direction as well as intensity. Since ants spend their life in contact with the ground, the soil surface makes a good place to leave a pheromone trail that can be followed by other ants. In those species which forage in groups, when a forager finds food they mark a trail on the way back to the colony, and this is followed by other ants that reinforce the trail when they head back to the colony. When the food is exhausted, no new trails are marked by returning ants and the scent slowly dissipates. This behavior helps ants adapt to changes in their environment. When an established path to a food source is blocked by a new obstacle, the foragers leave the path to explore new routes. If successful, the returning ant leaves a new trail marking the shortest route. Successful trails are followed by more ants, and each reinforces the trail with more pheromone (ants will follow the heaviest marked trails). Home is often located by remembered landmarks in the area and by the position of the sun; ants' compound eyes have specialized cells that detect polarized light, used to determine direction.

Ants use pheromones for other purposes as well. A crushed ant will emit an alarm pheromone which in high concentration sends nearby ants into an attack frenzy; and in lower concentration, merely attracts them. To confuse enemies, several ant species use "propaganda pheromones," which cause their enemies to fight amongst themselves.

Pheromones are produced by a wide range of glandular structures including cloacal glands, Dufour's glands, the hindgut, poison glands, pygidial glands, rectal glands, sternal gland and tibial glands on the back legs.

Pheromones are also exchanged mixed with food and passed in the trophallaxis, giving the ants information about one another's health and nutrition. Ants can detect what task group (e.g. foraging or nest maintenance) other ants belong to. When the queen stops producing a specific pheromone the workers raise new queens.

Some ants also produce sounds by stridulation using the gaster segments and also using their mandibles. They may serve to communicate among colony members as well as in interactions with other species.

13 For a relatively clear description of how the débâcle happened see Jeff Faux (a wonderfully appropriate name for his topic), "Is This the Big One?" in *The Nation*, 266: 14 (April 14, 2008), 11–12: "In a typical deal, subprime mortgages were sold to investment companies, where they were commingled with prime mortgages to back up new securities that could be touted as both safe and high-yielding. This new debt paper [Probably no paper was involved, just computer files. Faux uses an old-fashioned and now false term. I have never laid eyes on the actual paper versions of the stocks and bonds my broker tells me I own. I doubt if any such things exist. Calling it 'debt paper' is a

little like speaking, as everyone does, of "the banking industry," as if banks produced something solider than money out of money. (JHM)] was then peddled to investors, who used it as collateral for 'margin' loans to buy yet more stocks and bonds. At each change of hands, fees and underwriting charges added to the total claims on the original shaky mortgages. The result was a frenzied bidding up of prices for a bewildering maze of arcane securities that neither buyers nor sellers could accurately value.

Giant Ponzi scheme? Not to worry, responded the Wall Street geniuses. By spreading risks among more people, the miracle of 'diversity' was actually turning bad loans into good ones. Anyway, banks were buying insurance policies against default, which in turn were transformed into a set of even murkier securities called 'credit default swaps' and marketed to hedge funds, pension managers and in some cases back to the banks that were being insured in the first place. At the end of 2007 the market for these swaps was estimated at $45.5 trillion—roughly twice as large as all US stock markets combined." "Change of hands," a.k.a. "sleight of hand," "trans-formed," "swaps": Faux's language matches Freud's or Derrida's terminology of *Umsetzungen*, *Übersetzungen*, and *Übertragungen*, transfers, translations, metaphors, substitutions. Faux's terms, though they name computer work, also contain scarcely submerged metaphorical references to the techniques of magicians and mediums. For a more recent lucid description of what has happened, with a focus on Citigroup, see Thomas Friedman, 'All Fall Down', *New York Times*, November 26, 2008 (http://www.nytimes.com/2008/11/26/opinion/26friedman.html?th&emc=th).

14 See Danny Schechter, "Market Media Meltdown," in *The Nation*, 266:20 (May 26, 2008), 6, 8, for an account of the media's complicity.

15 See: http://jenson.stanford.edu/uhtbin/cgisirsi/wELISSqge4/GREEN/118050113/9. For a fascinating discussion of the development of the phonograph in the late nineteenth century and early twentieth century, along with its appearance in literature, and its early use to record the voices of poets like Browning and Tennyson, as well as politicians like Gladstone, see chapter 4, "The Recorded Voice: From Victorian Aura to Modernist Echo," in John M. Picker, *Victorian Soundscapes* (Oxford: Oxford University Press, 2003), 110–45. As Picker observes, early gramophone recordings were often made of distinguished people in their old age. Such recordings were

seen as a way to preserve the ghostly voices of the dead. Browning's fragmentary recording of a few lines from "How They Brought the Good News from Ghent to Aix" (he forgot his own poem after three and a half lines) was played on the anniversary of the poet's death and then again on the anniversary of his funeral, "mixing technology and spiritualism" (Picker, *op. cit.*, 123). James Joyce, in the "Hades" section of *Ulysses*, has Bloom imagine a family playing the recorded voice of their dead great grandfather: "Besides how could you remember everybody? Eyes, walk, voice. Well, the voice, yes: gramophone. Have a gramophone in every grave or keep it in the house. After dinner on a Sunday. Put on poor old greatgrandfather Kraahraark! Hellohellohello amawfullyglad kraark awfullygladaseeragain hellohello amarawf kopthsth. Remind you of the voice like the photograph reminds you of the face" (http://www.online-literature. com/view.php/ulysses/6?term=grand father). Steven Conner, echoing Joyce, connects the gramophone with ghosts and spiritualism in "A Gramophone in Every Grave," in his *Dumbstruck: A Cultural History of Ventriloquism* (Oxford: Oxford University Press, 2000), 362–93. The now classic studies of the cultural effects of new communications technologies during the period from before Browning on to Freud are Friedrich Kittler, *Discourse Networks 1800/1900*, trans. Michael Metteer, with Chris Cullins (Stanford, California: Stanford University Press, 1990), and *ibid.*, *Gramophone Film Typewriter*, trans. Geoffrey Winthrop-Young and Michael Wutz (Stanford, California: Stanford University Press, 1999). John Picker footnotes many other works about the gramophone and the telephone, for example Avital Ronnell's admirable *The Telephone Book: Technology—Schizophrenia—Electric Speech* (Lincoln: University of Nebraska Press, 1989).

16 http://www.lib.msu.edu/vincent/samples.html.

17 *Alfred, Lord Tennyson Reads from His Own Poems*, introduced by Sir Charles Tennyson, Craighill LP,TC 1: CRS audiocassette, CR 9000.

18 Robert Browning, *Complete Poetical Works*, ed. Horace E. Scudder (Boston: Houghton Mifflin, ©1895), 400–401, henceforth CPW, followed by the page number.

19 William Shakespeare, *Hamlet*, V, 1: "By the Lord, Horatio, this three years I have took note of it, the age is grown so picked that the toe of the peasant comes so near the heel of the courtier he galls his kibe."

20 See, for example, Jacques Derrida, "Artifacualities," in *Echographies of*

Television: Filmed Interviews, Bernard Stiegler co-author, trans. Jennifer Bajorek (Cambridge, England: Polity Press; Malden, Massachusetts: Blackwell, 2002), 1–27; *ibid.*, "Artifactualités," in *Échographies: de la télévision: Entretiens filmés*, Bernard Stiegler co-author (Paris: Galilée/Institut national de l'audiovisuel, 1996), 11–35.

21 Lewis Carroll, *Through the Looking-Glass*, in *Alice in Wonderland*, ed. Donald J. Gray, Norton Critical edition (New York: W. W. Norton & Company, Inc., 1971), 150.

22 "The Wheel," ll. 7–8, from *The Tower* (1928), W. B. Yeats, *The Poems*, new ed., ed. Richard J. Finneran (New York: Macmillan, 1983), 211. Freud's *Beyond the Pleasure Principle* was published in 1920. It argues for a death-wish in all living things, including human beings, that co-opts the pleasure principle. "The organism wishes to die only in its own fashion," says Freud. "Thus these guardians of life, too, were originally the myrmidons of death" (SE 18: 39).

23 Royle says that telepathy is "closely linked to the so-called decline of Christianity in European and North American culture: a belief in telepathy, in the late nineteenth century, often (though by no means always) appears to have provided a kind of substitute for a belief in God" (*After Derrida* [Manchester: Manchester University Press, 1994], 72).

24 See *A Taste for the Secret*, with Maurizio Ferraris, trans. from the French and Italian by Giacomo Donis, ed. Giacomo Donis and David Webb (Cambridge: Polity, 2001), 71. First published in Italian as *Il Gusto del Segreto* (Roma-Bari: Gius. Laterza and Figli Spa, 1997).

25 See, for example, http://www.nytimes.com/2009/01/30/ us/ 30suicide.html?scp=1&sq=veterans%27+suicide&st=nyt.

26 Carroll, *op. cit.*, 163.

27 http://www.marxists.org/archive/marx/works/1867-c1/ch01.htm#S4.

28 Jacques Derrida, *Archive Fever: A Freudian Impression*, trans. Eric Prenowitz (Chicago: The University of Chicago Press, 1996), 16, henceforth AF, followed by the page number; *ibid., Mal d'Archive* (Paris: Galilée, 1995), 33, henceforth MA, followed by the page number.

29 http://www.nytimes.com/2008/06/06/world/middleeast/06intel. html?th&emc=th.

30 Someone from the audience in an oral presentation of these remarks

astutely observed that my choice of an Australian aboriginal weapon as a figure resonates with Freud's famous appropriation of the concept of a "fetish" from the anthropology of his time. I had in mind the child's version of a boomerang that, if thrown correctly, will return to the hand of the thrower, whereas a real boomerang is a weapon used for catching birds by driving them into a net or for killing kangaroos and emus, though if it misses it will return to the thrower's hand, just as the child's toy does. If you miss catching it, you may get hurt. (http://en.wikipedia.org/wiki/Boomerang)

31 See Jacques Derrida, "Circumfession," in Bennington and Derrida. *Jacques Derrida*, trans. Geoffrey Bennington (Chicago: University of Chicago Press, 1993), 159; "*Circonfession*," in Geoffrey Bennington and Jacques Derrida, *Jacques Derrida* (Paris: Seuil, 1991), 150.

32 Jacques Derrida, *Glas*, trans. John P. Leavey, Jr. and Richard Rand (Lincoln, Nebraska: University of Nebraska Press, 1986), 196–7b; *ibid., Glas* (Paris: Galilée, 1974), 219b. This fragment of a poem Derrida published at the age of seventeen is cited on Valerio Adami's admirable Derrida lithograph and poster. The lithograph is reproduced in Adami's collection, *le voyage du dessin*, in *Derrière le miroir*, Number 214 (Paris: Maeght Editeur, 1975). Derrida comments on the poem fragment in "+ *R (par dessus le marché)*" in the same issue of *Derrière le miroir*. The pages are not numbered.

33 I borrow this wonderfully apt passage equating occultism with sludge from Pamela Thurschwell's citation of it (*op. cit.*, 120–1).

34 This information is taken from Daniel Karlin's notes to the poem in Robert Browning, *Selected Poems*, ed., with introduction and notes, Daniel Karlin (London: Penguin, 2004), 324.

35 These details, and others in subsequent paragraphs, come from W. C. DeVane, *A Browning Handbook*, 2nd ed. (New York: Appleton-Century-Crofts, Inc., ©1963), 307–12.

36 http://www.the-artists.org/posters/posters.php?item=1872710

37 *Études Anglaises*, VII (1954), 164.

38 See "The Impossible Profession," trans. Patience Moll, in Sarah Kofman, *Selected Writings* (Stanford, California: Stanford University Press, 2007), 56–70.

39 I long ago discussed the onomatopoeic grotesquerie or sludge of Browning's "style," halfway between solid and liquid, in the context of an extended reading of all Browning's poetry. See my "Robert Browning," in *The Disapperance of God* (Cambridge, Mass.: The

Belknap Press of Harvard University Press, 1963), 81–156. Browning's "Sibrandus Schnafnaburgensis" (CPW, 167) is a wonderful example of Browning poetry of the primeval soup, teaming with potential life.

40 See Karlin's note to the Bridgewater Treatises in the Penguin Browning, ed. cit., 328: "The Reverend Francis, Earl of Bridgewater (1758–1829), left money in his will for the writing of essays 'On the Power, Wisdom, and Goodness of God, as Manifested in the Creation.' The 'Bridgewater Treatises' were published 1833–40; the first, by Thomas Chalmers (1780–1847), was called 'The Adaptation of External Nature to the Moral and Intellectual Constitution of Man.'"

41 The title remains in French in the English translation. This essay is included in *The Post Card* (PC, 413–96; CP, 441–524).

42 The last words of *Beyond the Pleasure Principle* are a citation from a version by Rückert of one of the Maqâmât of al-Hariri: "*Was man nicht erfliegen kann, muss man erhinken. . . . Die Schrift sagt, es ist keine Sünde zu hinken.*" ("What we cannot reach flying we must reach limping . . . . The Book tells us it is no sin to limp.") (SE, 18: 64)

43 Here is the passage in de Man's "Sign and Symbol in Hegel's *Aesthetics*: "Hegel can therefore write the following quite astonishing sentence: 'When I say "I," I *mean* myself as *this* I to the exclusion of all others; but what I say, I, is precisely anyone; any I, as that which excludes all others from itself [*ebenso, wenn ich sage, "Ich," meine ich mich als diesen alle anderen Ausschließenden; aber was ich sage, Ich, ist eben jeder*]" (*Aesthetic Ideology*, ed. Andrzej Warminski [Minneapolis: University of Minnesota Press, 1996], 98).

44 Carroll, *op. cit.,* 94.

45 *Ibid.,* 94–5.

46 Aristotle, *Poetics*, 1453b18–24, trans. Gerald F. Else (Ann Arbor: The University of Michigan Press, 1978), 41. I owe this reference to a fine essay on hospitality by James Heffernan.

47 See PC, 165; CP, 179. This page contains a brief proleptic summary of what was to become Derrida's essay, "Des tours de Babel." The latter had not, I believe, yet been written in 1978.

48 See, for one example out of many, Jacques Derrida, *On Touching—Jean-Luc Nancy*, trans. Christine Irizarry (Stanford: Stanford University Press, 2005), 176; *ibid., Le toucher, Jean-Luc Nancy* (Paris: Galilée, 2000), 202.

49 Jacques Derrida, *Learning to Live Finally: An Interview with Jean Birnbaum*, trans. Pascale-Anne Brault and Michael Naas (Hoboken, New Jersey: Melville House Publishing, 2007); *ibid., Apprendre à vivre enfin* (Paris: Galilée, 2005).

50 Ernest Jones, *The Life and Work of Sigmund Freud*, ed. and abridged by Lionel Trilling and Steven Marcus (Garden City, N.Y.: Doubleday Anchor, 1963).

51 In *Beyond the Pleasure Principle* (SE, 18: 14–15), Freud wrote:

This good little boy, however, had an occasional disturbing habit of taking any small objects he could get hold of and throwing them away from him into a corner, under the bed, and so on, so that hunting for his toys and picking them up was often quite a business. As he did this he gave vent to a loud, long-drawn-out "o-o-o-o," accompanied by an expression of interest and satisfaction. His mother and the writer of the present account were agreed in thinking that this was not a mere interjection but represented the German word "*fort*'" ["gone"]. I eventually realized that it was a game and that the only use he made of any of his toys was to play "gone" with them. One day I made an observation which confirmed my view. The child had a wooden reel with a piece of string tied around it. It never occurred to him to pull it along the floor behind him, for instance, and play at its being a carriage. What he did was to hold the reel by the string and very skillfully throw it over the edge of his curtained cot, so that it disappeared into it, at the same time uttering his expressive "o-o-o-o." He then pulled the reel out of the cot again by the string and hailed its reappearance with a joyful "*da*" ["there"]. This, then, was the complete game—disappearance and return. As a rule one only witnessed its first act, which was repeated untiringly as a game in itself, though there is no doubt that the greater pleasure was attached to the second act.

The interpretation of the game then became obvious. It was related to the child's great cultural achievement—the instinctual renunciation (that is, the renunciation of instinctual satisfaction) which he had made in allowing his mother to go away without protesting.

I found this passage on the Internet, then cut and pasted it, in thirty seconds, with Google's help. There is postmodern scholarship for you, with the new medium as the maker! I did, however, check the passage against SE, 18, and made a few trivial corrections.

52 For a Freud family-tree see: http://www.thecjc.org/ftree.htm .

53 This was written before the United States presidential election of November 2008. With the election of Barack Obama we now (February 9, 2009) have the audacity of hope.

54 http://www.nytimes.com/2008/03/24/business/media/ 24murdoch. html?th&emc=th.

Index